Morphological Diagnosis of Humoral Cells under Light and Electron Microscopy

Dayong Gu • Ying Wang
Editors

Morphological Diagnosis of Humoral Cells under Light and Electron Microscopy

Springer

Editors
Dayong Gu
Laboratory Department
Shenzhen Second People's Hospital
Shenzhen, Guangdong, China

Ying Wang
Laboratory Department
Shenzhen Second People's Hospital
Shenzhen, Guangdong, China

ISBN 978-981-95-8667-7 ISBN 978-981-95-8668-4 (eBook)
https://doi.org/10.1007/978-981-95-8668-4

This Springer imprint is published by the registered company Springer Nature Singapore Pte Ltd.
The registered company address is: 152 Beach Road, #21-01/04 Gateway East, Singapore 189721, Singapore

Foreword

The publication of *Morphological Diagnosis of Humoral Cells: Light Microscopy & Electron Microscopy* was unexpected, as this is a basic clinical work. The inspiration for this book came from Chief Editor Wang Ying, who, while undergoing training at Zhejiang Provincial People's Hospital, was inspired by the translation of foreign morphology atlases and hoped that a similar atlas comparing light microscopy and electron microscopy could be created for fluid cell morphology. After hearing her idea, I was excited as well, as this would fill a gap in fluid cell morphology in China, helping colleagues recognize the ultrastructure of cells and explore cellular mysteries further. The compilation of such a book was indeed necessary.

However, the preparation and imaging of electron microscopy samples require substantial funding. Thanks to the strong support, sponsorship, and guidance from Professor Gu Dayong, Director of the Laboratory Department at Shenzhen Second People's Hospital, the project team successfully launched this plan. For more than six months, the team meticulously searched for various specimens, collecting them together with several trainees and clinical laboratory teachers from surrounding hospitals in Hangzhou. With the strong assistance of Dr. Li Junying from the Electron Microscopy Laboratory at the Zhejiang University Agrobiology and Environmental Test Center, we finally completed the preparation and imaging of over 110 electron microscopy samples. Despite the busy work schedules and the demands of their studies, the team members dedicated themselves to preparing, observing, and identifying cell types in the electron microscopy lab, while also diligently reviewing literature and organizing images and written materials. Their dedication and practical spirit deeply moved me.

"Morphological Diagnosis of Humoral Cells under Light and Electron Microscopy" is an excellent diagnostic reference book. Before its publication, there was no known development and application of electron microscopy in fluid cell morphology. The book adds an academic "feast" to basic fluid cell morphology, deepening our understanding of the ultrastructure of fluid cells. I believe this book will be well-received by laboratory personnel and medical students.

Zhejiang Provincial People's Hospital
Zhejiang, China

Wu Mao

Introduction

The application of electron microscopy (EM) in fluid cell morphology compensates for the limitations of conventional examinations in recognizing details and plays an essential role in tumor diagnosis, as well as in the diagnosis and analysis of complex cases. Currently, EM is primarily used for pathological tissue research in both domestic and international laboratories, with relatively few studies on fluid cell electron microscopy and even fewer on blood cell electron microscopy. Therefore, we hope to compile a monograph introducing fluid cell morphology through a comparative approach between light microscopy (LM) and EM.

We collected cases from multiple medical centers and conducted related morphological studies; it took us 2 years to complete this book after extensive research and gathering of information. Due to the wide range of sources, there may be errors or inaccuracies in the content, and we sincerely welcome feedback and criticism from our colleagues to help us improve in future editions.

We extend our deepest gratitude to Professor Wu Mao, who has continuously supported and helped us during the compilation of this book; to Professor Li Junying from the Electron Microscopy Laboratory at the Zhejiang University Agrobiology and Environmental Test Center; to Wang Weilan for their support in providing specimens and cases; and to Professor Zhou Yuli from the First People's Hospital of Hangzhou, Zhejiang University School of Medicine, Dr. Lu Xiaoya from Zhejiang Provincial People's Hospital, Lu Yu from Jiaxing First Hospital; Professor Cao Yu from the Affiliated Hospital of Zunyi Medical University, Qiu Qinghua from Panzhihua Integrated Chinese and Western Medicine Hospital, and Professor Lou Hongping from Hangzhou Normal University Affiliated Hospital for their support in the experimental operations. Special thanks to Professor Ru Jinwei from Lechang People's Hospital for drawing the diagrams and reviewing the manuscripts, and to Director Gu Dayong from the Laboratory Department of Shenzhen Second People's Hospital for his strong support. It is through the unity and concerted efforts of everyone that this book was successfully published.

"Even a thousand miles begins with a single step." We hope that the publication of this book will serve as a catalyst, attracting more experts in morphology to contribute to making EM morphology richer and more multidimensional, adding new perspectives to morphological diagnostics, and contributing to the development of morphology in China.

Shenzhen Second People's Hospital Wang Ying
Shenzhen, China

Editorial Board Members

Chief Reviewers
Wu Mao, Zhejiang Provincial People's Hospital
Xu Yong, Shenzhen Second People's Hospital

Chief Editors
Gu Dayong, Shenzhen Second People's Hospital
Wang Ying, Shenzhen Second People's Hospital

Associate Editors
Zhou Yuli, The First People's Hospital of Hangzhou, Zhejiang University School of Medicine
Ru Jinwei, Lechang People's Hospital
Li Junying, Zhejiang University
Lu Xiaoya, Zhejiang Provincial People's Hospital
Cao Yu, The Affiliated Hospital of Zunyi Medical University
Li Yanwu, Shenzhen Second People's Hospital
Ding Fanglin, Shenzhen Second People's Hospital

Contributors (in order of surname strokes)
Lu Yu, Jiaxing First Hospital
Qiu Qinghua, Panzhihua Integrated Chinese and Western Medicine Hospital
Zhang Bo, PLA 83rd Group Army Hospital
Zheng Liheng, Hebei Chest Hospital
Lou Hongping, Hangzhou Normal University Affiliated Hospital

Translators List

Consultant Dayong Gu Mao Wu

Lead Translator
Ying Wang, Shenzhen Second People's Hospital
Guangtao Huang, Shenzhen Second People's Hospital

Deputy Lead Translator
Ningchao Du, Shenzhen Second People's Hospital
Bo Situ, Nanfang Hospital, Southern Medical University

Contributor
Chuanghua Qiu, Shenzhen Second People's Hospital
Xiaoqing Li, Shenzhen Second People's Hospital
Xingjun Meng, Women' Hospital, School of Medicine, Zhejiang University
Yu Lu, The First Hospital of Jiaxing
Shiping He, Shenzhen Second People's Hospital
Yufan Wu, Shenzhen Second People's Hospital
Lijuan Xiang, Shenzhen Second People's Hospital
Jinwei Ru, Lechang People's Hospital
Hao Deng, Shenzhen Second People's Hospital
Tiantian Lu, Shenzhen Second People's Hospital
Lin Lin, Shenzhen Second People's Hospital
Yongbo Nie, Shenzhen Second People's Hospital
Yong Luo, Shenzhen Second People's Hospital

Summary

The application of electron microscopy (EM) in fluid cell morphology complements the limitations of conventional examinations in recognizing details, playing a crucial role in the diagnosis and differential diagnosis of tumors and complex cases. Starting with basic cell biology, the editors analyze cellular structures and morphological characteristics under EM, comparing them with light microscopy (LM) images to establish a link between LM and EM morphologies. Through real case studies, they compare the abnormal cell features observed in both LM and EM, explaining the reasons behind the morphological changes seen under LM based on the ultrastructural characteristics observed under EM. This approach helps laboratory physicians grasp the essence behind the phenomena, making it easier to understand the changes in morphological characteristics. The book is concise in its explanations, rich in images, and adopts a comparative approach between LM and EM to meticulously introduce the relevant knowledge of fluid cell morphology. The book is of great reference value for understanding the ultrastructure of fluid cells and is recommended for clinical laboratory professionals, university faculty, and students.

Contents

About the Chief Reviewers

Wu Mao is a licensed physician and Chief Technologist (Professor Level III). He currently holds the positions of Professor and Deputy Director of the Fundamental Teaching and Research Office of Clinical Laboratory Science at Hangzhou Medical College. He previously served as Director of the Cytology Laboratory at the Clinical Testing Center of Zhejiang Provincial People's Hospital. Dr. Wu is the Chief Cytomorphology Expert at ADICON Clinical Laboratories and Chairman of the Management Committee of the Jun'an Medical Cell Platform. His additionalprofessional and academic appointments include Member of the Laboratory Medicine Branch of the National Vocational Education Teaching Steering Committee, Member of the Expert Committee of the Yangtze River Delta Morphological Diagnosis Alliance, Member of the Clinical Laboratory Professional Group of the China Association of Equipment, and Member of the Laboratory Medicine Branch of the Zhejiang Medical Doctor Association.His primary expertise encompasses cytological techniques and laboratory diagnostics, specializing in bone marrow and peripheral blood cell analysis, cytology of body fluids (including serous cavity effusions), and fine-needle aspiration cytology. Dr. Wu founded the WeChat public account "Wu Mao Talks Cells" and established the "Jun'an Medical Cell Platform" to facilitate professional education and exchange in the field. He has led five scientific and technological achievements recognized with awards, including three Second Prizes and one Third Prize of the Zhejiang Provincial Medical and Health Science and Technology Innovation Award, as well as one Third Prize of the Zhejiang Provincial Science and Technology Progress Award. In addition, he has authored five books, served as chief reviewer for five publications, co-edited three university-level textbooks, led the drafting and formulation of three Chinese expert consensus documents, and published over 40 academic papers.

Xu Yong holds an M.D. degree and completed postdoctoral training. He currently serves as the Secretary of the Party Committee of Shenzhen Third People's Hospital. He also holds a Master of Hospital Administration (MHA) from Flinders University, Australia. Dr. Xu is Director of the Shenzhen Key Laboratory of Molecular Diagnostics in Medical Laboratory Science, and serves as a Master's Supervisor in Medical Laboratory Science at the School of Medicine, Jinan University, as well as a part-time Master's Supervisor at the School of Public Health, Wuhan University, and the School of Medicine and Health Management, Tongji Medical College, Huazhong University of Science and Technology. In addition, he is Chair of the Integrated Medicine Committee of Shenzhen Medical Association, Vice President of Shenzhen Hospital Administrators Association and Chair of its Operations and Evaluation Committee, Executive Council Member of Shenzhen Medical Association, and Editorial Board Member of the Modern Hospital Management Journal. Dr. Xu has led and completed multiple research projects, including a subtask under a National Key R&D Program, the Shenzhen Key Laboratory of Medical Laboratory Science, two research projects supported by Guangdong Provincial Health Department, six by Shenzhen Science and Technology Bureau, one by Shenzhen Science and Technology Innovation Commission, and one Shenzhen Key Research Project. He has also participated in one provincial key research program. To date, he has published over 70 SCI-indexed papers in high-level domestic and international journals, supervised the graduation of six Master's students, contributed to one foreign-language academic monograph, and filed two national invention patents.

About the Chief Editors

Gu Dayong MD, is a Chief Physician and a leading high-level scientific and technological talent in Shenzhen. He currently holds several prominent academic and professional positions, including Executive Council Member of the China Society for Inspection and Testing, Member of the National Technical Committee for Standardization of Biochemical Testing (SAC/TC387), Standing Committee Member of the Primary Care Laboratory Technology Standardization Branch of the China Association for the Promotion of Medical and Health Care, Vice Chairman of the Guangdong Tropical Medicine Society, Vice Chairman of the Sensor Technology Branch of the Guangdong Society of Biomedical Engineering, Standing Committee Member of the Laboratory Medicine Branch of the Guangdong Medical Doctor Association, Member of the Laboratory Medicine Branch of the Guangdong Medical Association, Vice Chairman of the Laboratory Medicine Branch of the Shenzhen Medical Association, Vice President of the Laboratory Medicine Branch of the Shenzhen Medical Doctor Association, and Vice Chairman of the Integrated Medicine Committee of the Shenzhen Hospital Administrators Association.With over 30 years of experience in clinical laboratory diagnostics and related research, Dr. Gu has led more than 30 research projects, including those under the National Key R&D Program, the National Natural Science Foundation of China, and various provincial and municipal initiatives. He has been granted over 30 PCT and invention patents, received six provincial- and ministerial-level science and technology awards, and contributed to the publication of 12 national and industry standards. To date, he has authored more than 100 academic papers, including over 40 indexed in SCI, and has served as chief editor or translator of three professional monographs.

Wang Ying is an Associate Chief Technologist and Physician. She graduated from the Department of Medical Laboratory Science at the Third Military Medical University and currently works in the Clinical Laboratory Department of Shenzhen Second People's Hospital. With over 20 years of experience in laboratory medicine, she has accumulated extensive clinical expertise. Additionally, she has served as a lecturer in Laboratory Diagnostics at Shenzhen University for 11 years. Her professional roles include Member of the Clinical Laboratory Committee under the Laboratory Medicine Branch of Guangdong Medical Association, Member of the Neuro-Ophthalmology Committee of Shenzhen Optometry Society, and Member of the Laboratory Branch of Guangdong Health Care Association. Driven by a strong dedication to morphological studies, she has undertaken multiple advanced training programs in morphology: in 2011, she studied genetics at Peking University Shenzhen Hospital; in 2016, she trained in hematopathology morphology, histochemistry, coagulation, and hemolysis at the Institute of Hematology & Blood Diseases Hospital, Chinese Academy of Medical Sciences; and in 2021, she completed further training in body fluid cytomorphology at Zhejiang Provincial People's Hospital. She has won awards in eight national, provincial, and municipal morphology competitions, participated in multiple research projects, filed 21 invention patents as the first inventor, served as chief editor or contributing author of three professional books, and has published numerous academic papers.

Overview

1

Dayong Gu and Ying Wang

1.1 Development of Cellular Microscopy Techniques

Cells are the basic units of life, and due to their extremely small size, they cannot be directly observed with the naked eye. It was not until the invention of the microscope in the seventeenth century that cell observation became possible. The light microscope (LM) uses visible light to magnify objects through an optical magnification system (eyepiece and objective lens) by several hundred times for observation. It can be directly used to observe unicellular organisms or cultured cells in vitro and observe the structures and morphology of cells after fixation and staining. The resolution can reach 0.2 μm, much finer than the resolution of the naked eye (0.2 mm), and is convenient to use, making it the most widely used equipment for microscopic observation today. In addition to the commonly used ordinary light microscope, there are many different types of optical microscopes, such as phase contrast microscopes (distinguishing areas with different densities), differential interference contrast microscopes (observing the movement of living cell particles and organelles), fluorescence microscopes (facilitating the identification of proteins labeled with different fluorophores and the tissue structures they form), laser scanning confocal microscopes (used for the localization and dynamic changes of subcellular structures and components), and super-resolution microscopy (including total internal reflection fluorescence microscopy, single-molecule imaging-based PALM/STORM methods, 4π microscopy, STED microscopy, and structured illumination microscopy).

Due to the wavelength limitation of light waves, the resolution of light microscopy (LM) is difficult to further improve, making it challenging to better distinguish and analyze some structures and morphologies of cells. Using an electron microscope (EM), which has a higher resolution, allows us to observe the ultrastructure inside cells. In the context of precision medicine, using EM to identify cells has

D. Gu (✉) · Y. Wang
Laboratory Department, Shenzhen Second People's Hospital, Shenzhen, Guangdong, China

D. Gu, Y. Wang (eds.), *Morphological Diagnosis of Humoral Cells under Light and Electron Microscopy*, https://doi.org/10.1007/978-981-95-8668-4_1

become a need and trend. We hope that this book will serve as an introduction, presenting the significance and clinical application of EM in cell morphology and exploring the broader space for EM in practical applications.

1.1.1 Differences Between Electron Microscopy and Light Microscopy

Electron microscopy uses electron beams as a light source, which has a much shorter wavelength than visible light, generally less than 0.1 nm. The difference in light sources determines a series of differences between EM and LM. The resolution of EM can reach 0.2 nm, and the magnification can be up to 10^6 times. Electron microscopy uses electron beams with wavelengths of 0.01–0.09 nm as a light source, and the lenses are electromagnetic lenses. The samples need to be observed in a vacuum, and the images are formed by the scattering and transmission of electrons through the samples, which are recorded on photographic film or a CCD camera (Fig. 1.1).

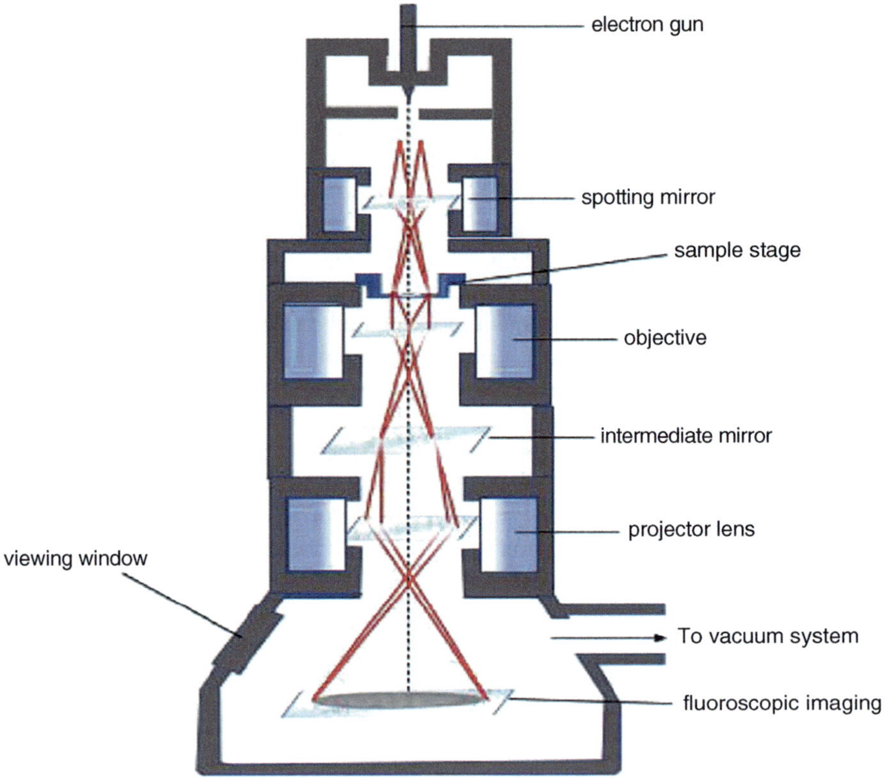

Fig. 1.1 Basic structure and imaging principle of electron microscope

1.1.2 Preparation of EM Samples

EM samples need to be prepared to meet imaging requirements, possibly using techniques such as ultrathin sectioning, negative staining, freeze-etching, 3D reconstruction, cryo-electron microscopy, and scanning electron microscopy. Currently, we mostly use transmission electron microscopy (TEM) and scanning electron microscopy (SEM). TEM requires the fixation, staining, embedding, and sectioning of samples to produce ultrathin sections for observing the ultrastructure inside cells under TEM; SEM observes directly after fixation and staining. Sample fixation is generally done with glutaraldehyde and osmium tetroxide (OsO4, commonly referred to as osmium acid), and the staining agents are heavy metals—osmium tetroxide stains lipids, lead citrate stains proteins, and uranyl acetate is used to stain nucleic acids. When the electron beam passes through the sample, metal ions scatter and absorb electrons to varying degrees, creating light and dark contrasts. Therefore, EM images are black and white. SEM samples only require fixation and staining, and the resulting images have a strong 3D effect, allowing the observation of the three-dimensional structure outside the cells.

1.2 Cellular Ultrastructure

Under an ordinary LM, after cell staining, only the cell membrane, cytoplasm, nucleus, nucleolus, and variously sized and colored granules and inclusions within the cytoplasm can be seen, but the membrane structures, organelles, and specific contents of the inclusions cannot be observed. In clinical diagnosis, it can be difficult to identify very similar cells, and EM, along with immunohistochemistry, is one of the confirmatory diagnostic methods that can be used to improve diagnostic capabilities and levels. Currently, although EM technology is widely used in scientific research, its application in clinical auxiliary diagnosis is limited. There are few atlases of blood cell ultrastructure, and no related references in the field of fluid cell cytology have been found. We hope this book can expand into a new field and better support clinical diagnosis.

First, we introduce the ultrastructure of cells to facilitate the understanding and analysis of the morphological differences of cells under different conditions observed through EM (Fig. 1.2).

1.2.1 Plasma Membrane

The plasma membrane, formerly known as the cell membrane, refers to the outermost layer of the cell, composed of lipids, proteins, and carbohydrates. As a boundary membrane, the plasma membrane plays a vital role in maintaining the relative stability of the cell, transporting substances and energy between cells, and transmitting information. It is collectively referred to as the biomembrane along with the internal membrane system (the membranes of various organelles) of the cell.

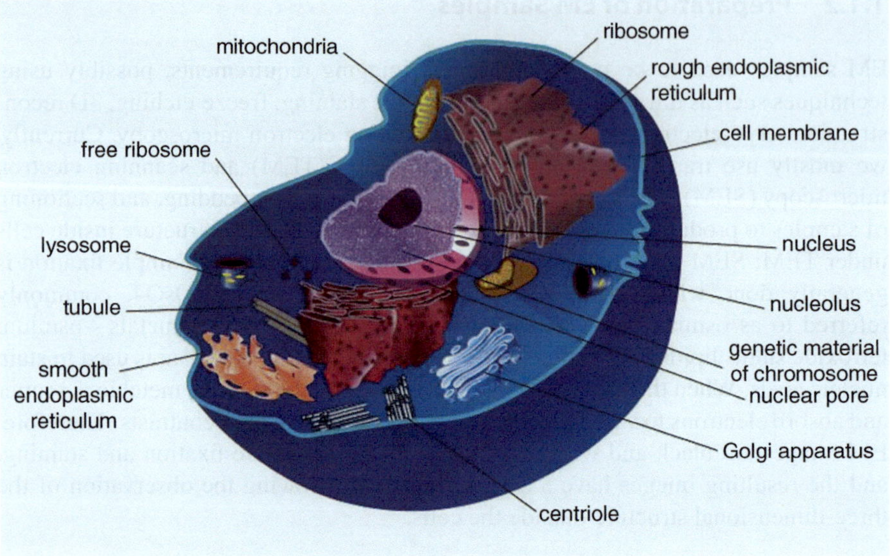

Fig. 1.2 Schematic diagram of cell ultrastructure
细胞膜——cell membrane；细胞核——nucleus；核仁——nucleolus；染色质——genetic material of chromosome；核孔——nuclear pore；高尔基体——Golgi apparatus；中心粒——centriole；滑面内质网——smooth endoplasmic reticulum；微管——tubule；溶酶体——lysosome；游离核糖体——free ribosome；线粒体——mitochondria

Endocytosis (Fig. 1.3) can be divided into phagocytosis and pinocytosis. Typically, the diameter of phagocytic vesicles formed by phagocytosis is greater than 250 nm, while the diameter of pinocytic vesicles formed by pinocytosis is less than 150 nm. All eukaryotic cells can continuously ingest solutions and soluble molecules through pinocytosis, while phagocytosis occurs mostly in specialized phagocytic cells (such as macrophages). Phagocytosis is a specialized form of endocytosis that requires the binding of the phagocytic material to receptors on the surface of phagocytic cells, triggering a cellular response—a signal-triggered process. Unlike phagocytosis, pinocytosis occurs in almost all types of eukaryotic cells. Pinocytosis can be further divided into clathrin-dependent endocytosis, caveolae-dependent endocytosis, macropinocytosis, and non-clathrin/caveolae-dependent endocytosis (Fig. 1.4). Macropinocytosis is similar to phagocytosis, as the endocytic vesicles formed are also large, but there are clear differences between the two, such as the receptors that initiate macropinocytosis, which are located on the surface of some types of cells and can initiate other physiological functions. Some receptors are related to cell growth factors, explaining why some non-macrophage cells can also phagocytose and form inclusions, such as mesothelial cells and tumor cells.

Exocytosis is the opposite of endocytosis; it is the process of expelling substances from the cell by merging secretory vesicles or other vesicles with the plasma membrane. This process is essential for material exchange, signal transmission,

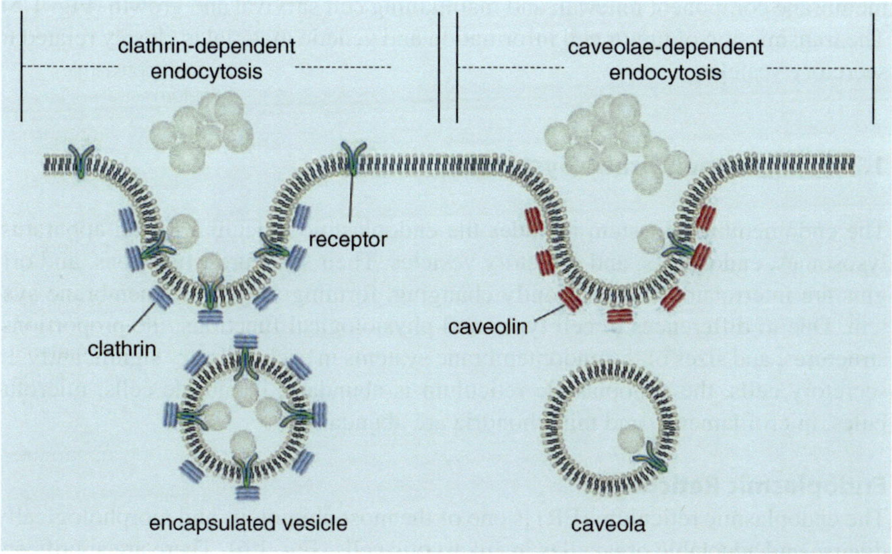

Fig. 1.3 Schematic diagram of clathrin-dependent endocytosis and caveolae-dependent endocytosis
网格蛋白依赖的胞吞作用—— clathrin-dependent endocytosis；胞膜窖依赖的胞吞作用——caveolae-dependent endocytosis；网格蛋白——clathrin；受体——receptor；窖蛋白——caveolin；包被膜泡——encapsulated vesicle；膜窖体——caveola

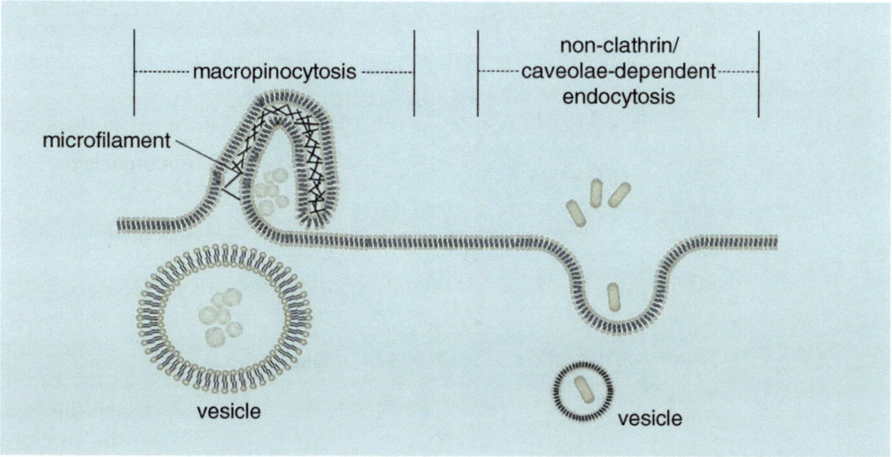

Fig. 1.4 Schematic diagram of macropinocytosis and non-clathrin/caveolae-dependent endocytosis
大型胞饮作用——macropinocytosis；非网格蛋白 / 胞膜窖依赖的胞吞作用——non-clathrin/caveolae-dependent endocytosis；微丝——microfilament；囊泡——vesicle；小泡——vesicle

membrane component renewal, and maintaining cell survival and growth (Fig. 1.5). The transmission of tumor cell information and genetic material is closely related to secretory vesicles.

1.2.2 Endomembrane System

The endomembrane system includes the endoplasmic reticulum, Golgi apparatus, lysosomes, endosomes, and secretory vesicles. Their structures, functions, and origins are interrelated and constantly changing, forming a dynamic membrane system. Due to differences in cell types and physiological functions, the proportions, structures, and sizes of the endomembrane systems in each cell vary significantly. In secretory cells, the endoplasmic reticulum is abundant; in muscle cells, microtubules, microfilaments, and mitochondria are abundant.

Endoplasmic Reticulum

The endoplasmic reticulum (ER) is one of the most ubiquitous and morphologically diverse and adaptable organelles in eukaryotic cells (Fig. 1.6). There are significant differences in the endoplasmic reticulum among different types of cells, and the structure and function of the ER in the same cell change constantly at different developmental stages. It synthesizes a series of macromolecules such as proteins,

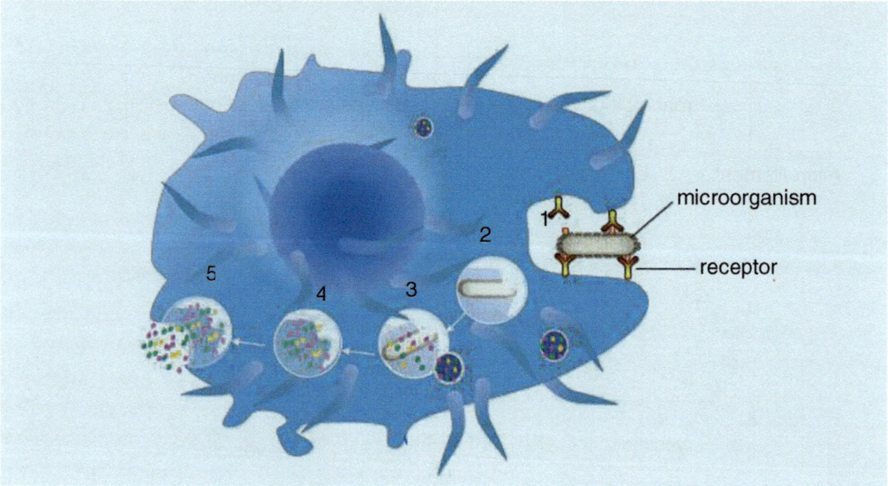

Fig. 1.5 Schematic diagram of exocytosis
(1) Binding and absorption; (2) phagocytosis; (3) Phagosomes and lysosomes form phagolysosomes; (4) digestion; (5) Release of degradation products
微生物——microorganism; 受体——receptor

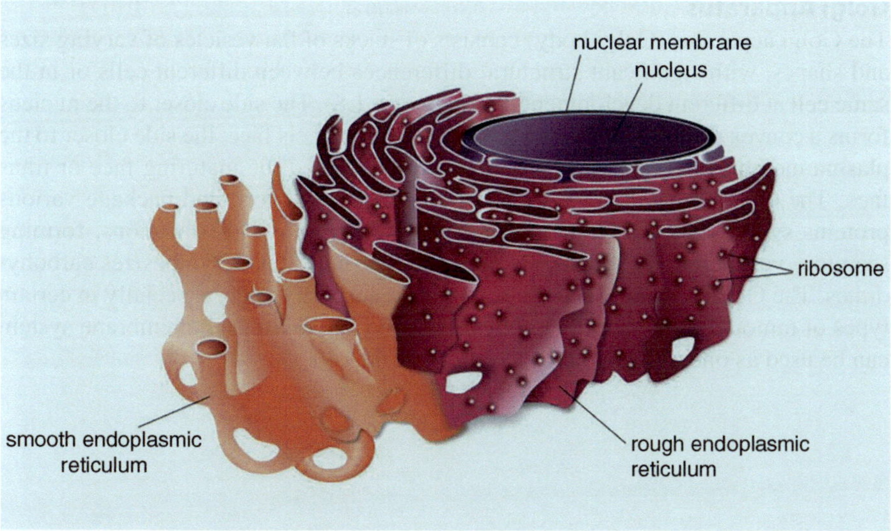

Fig. 1.6 Schematic diagram of endoplasmic reticulum
核膜——nuclear membrane；细胞核——nucleus；核糖体——ribosome；粗面内质网——rough endoplasmic reticulum；滑面内质网——smooth endoplasmic reticulum

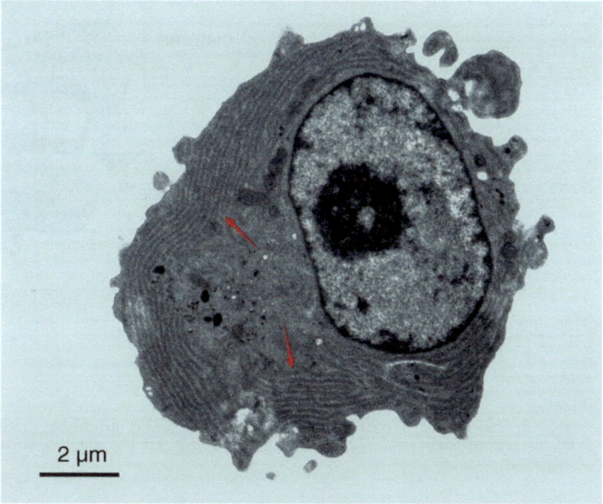

Fig. 1.7 Rough endoplasmic reticulum system of multiple myeloma cells with well-developed plate laminae (TM，12 000×)

lipids, and glycogen, except for nucleic acids. The ER attached with ribosomes is called rough ER, mainly involved in protein synthesis; smooth ER, lacking ribosomes, primarily synthesizes lipids and is common in cells secreting steroid hormones. The ER is abundant in cells with strong synthesis functions (Fig. 1.7) and is sparse in some undifferentiated or immature cells (such as primitive, immature cells, and undifferentiated tumor cells).

Golgi Apparatus

The Golgi apparatus (Golgi body) consists of stacks of flat vesicles of varying sizes and shapes, with significant structural differences between different cells or in the same cell at different developmental stages (Fig. 1.8). The side closer to the nucleus forms a convex shape and is called the forming face or cis face; the side closer to the plasma membrane forms a concave shape and is called the maturing face or trans face. The Golgi apparatus mainly functions to process, sort, and package various proteins synthesized by the ER, transporting them to specific locations, forming secretory vesicles, and releasing them outside the cell. It also synthesizes carbohydrates. The Golgi apparatus is very abundant in secretory cells, especially in certain types of tumor cells, where the highly developed and dense endomembrane system can be used as one of the diagnostic criteria (Fig. 1.9).

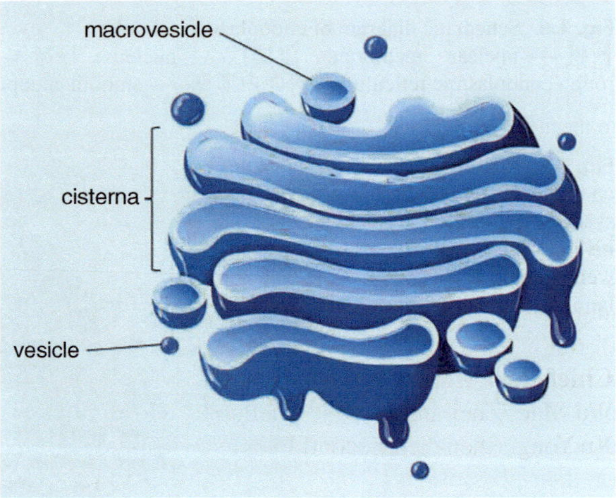

Fig. 1.8 Schematic diagram of Golgi apparatus 大囊泡——macrovesicle; 扁平囊——cisterna; 小囊泡——vesicle

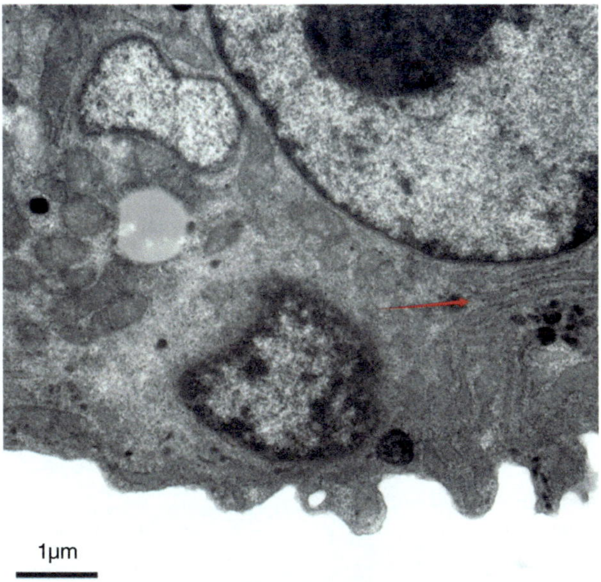

Fig. 1.9 Well-developed Golgi bodies in multiple myeloma cells (TM , 20 000×). Light microscopic Ritter staining does not reveal organelle structures, only pale stained areas next to the nucleus are visible

Lysosomes

Lysosomes (Fig. 1.10) are vesicular organelles surrounded by a single membrane and containing various acidic hydrolases. Their main function is to dissolve and digest ingested substances (or aging, damaged organelles) within the cell, playing an essential role in maintaining cellular metabolism and defense. The morphology and size of lysosomes vary according to the different enzymes; they contain and are classified into primary lysosomes, secondary lysosomes, and residual bodies based on their physiological function stages (Fig. 1.11). Secondary lysosomes are formed by the fusion of primary lysosomes with autophagic vesicles or heterophagic vesicles, which digest the contents, and are called autolysosomes and heterolysosomes, respectively. After digestion in secondary lysosomes, small molecules can be reused by the cell, while undigested materials remain within the lysosome as residual bodies, which can be expelled from the cell through exocytosis. The autophagy function of lysosomes plays an important role in the proliferation and metastasis of tumor cells, as it increases the reusability of substances within tumor cells, providing the necessary materials for their mass proliferation. Additionally, it can recycle organelles damaged during metastasis by fluid shear forces, helping maintain cellular stability. Genetic defects leading to the deficiency of specific lysosomal enzymes result in the accumulation of corresponding substrates within the cell, such as in Tay-Sachs disease, where the inability to degrade GM2 ganglioside effectively leads to lysosomal storage disease.

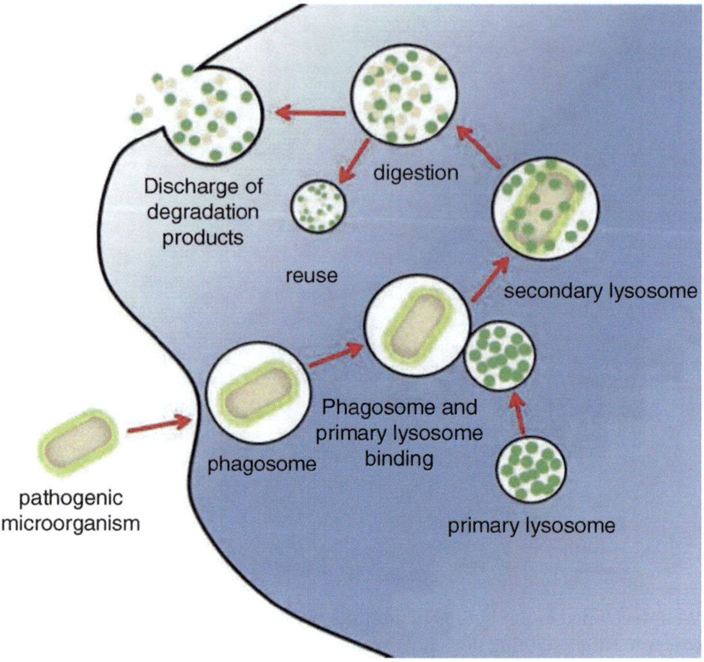

Fig. 1.10 Schematic diagram of Lysosomes
病原微生物——pathogenic microorganism；吞噬体——phagosome；初级溶酶体——primary lysosome；吞噬体和初级溶酶体结合——phagosome and primary lysosome binding；次级溶酶体——secondary lysosome；消化——digestion；再利用——reuse；降解物排出——discharge of degradation products

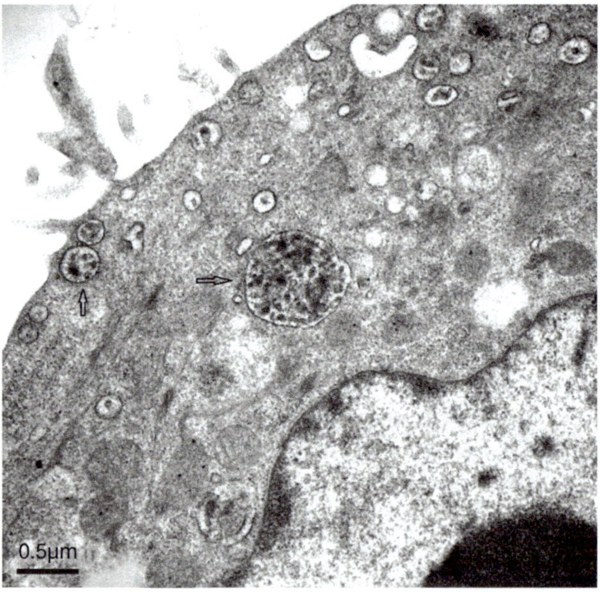

Fig. 1.11 Secondary lysosomes (hollow arrows) that have phagocytosed material; primary lysosomes are usually small and without inclusions (**TM，25 000×**)

1.2.3 Mitochondria

Mitochondria are special organelles in eukaryotic cells enclosed by a double membrane, responsible for energy production and conversion (Fig. 1.12). Their size and morphology vary significantly depending on the structure and function of the cell, but the basic structure is composed of inner and outer membranes. The outer membrane is flat and serves as a boundary, while the inner membrane folds inward, forming cristae. The space between the outer and inner membranes is called the intermembrane space, with a typically stable width. Mitochondria are abundant in cells with active synthesis and metabolism (Fig. 1.13) and can serve as a characteristic for identifying tumor cells.

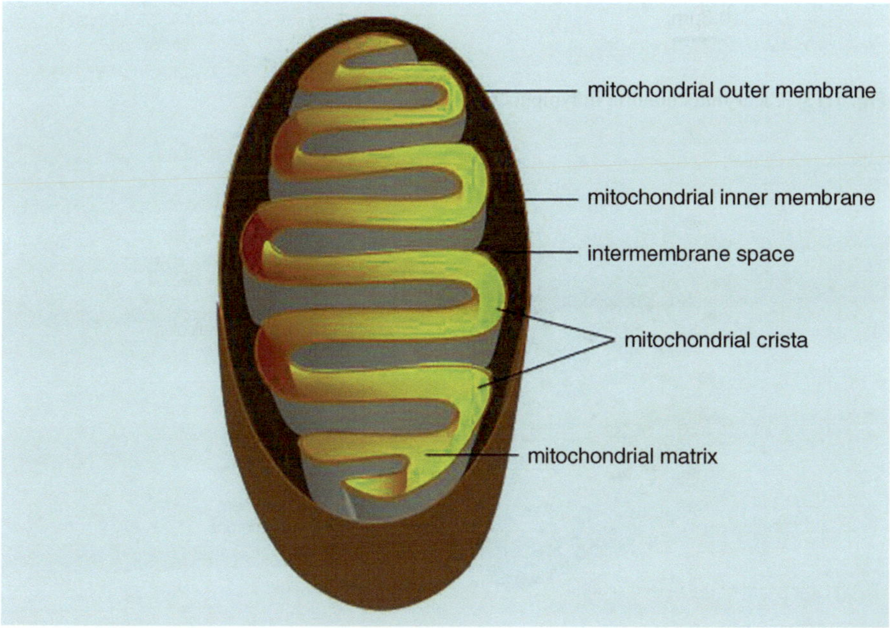

Fig. 1.12 Schematic diagram of Mitochondria
外膜——mitochondrial outer membrane; 内膜——mitochondrial inner membrane; 膜间隙——intermembrane space; 嵴——mitochondrial crista; 基质——mitochondrial matrix

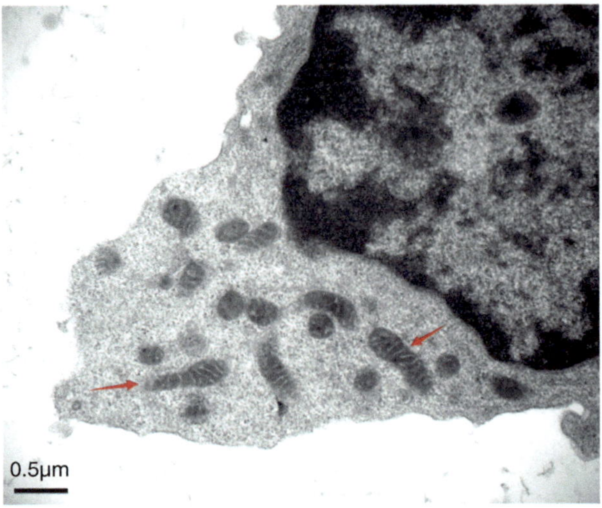

Fig. 1.13 Clear mitochondria in lymphocytes (**TM，30 000×**)

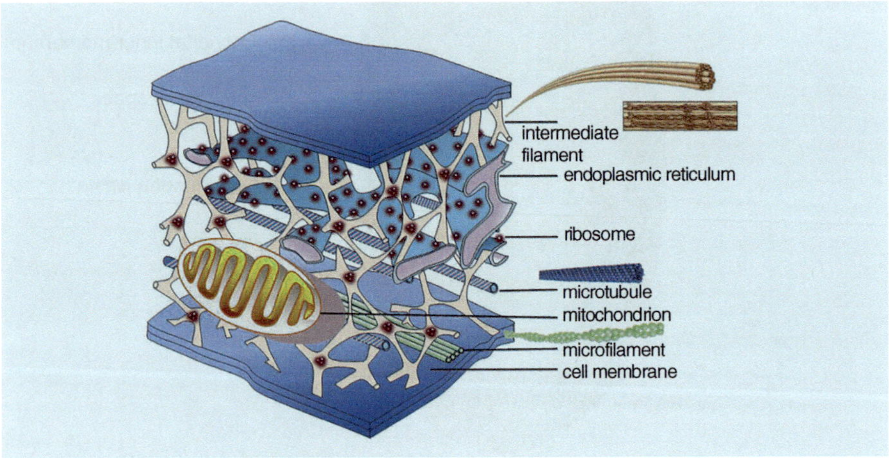

Fig. 1.14 Schematic diagram of the cytoskeleton
中间丝——intermediate filament；内质网——endoplasmic reticulum；核糖体——ribosome；微管——microtubule；线粒体——mitochondrion；微丝——microfilament ；细胞膜——cell membrane

1.2.4 Cytoskeleton

The cytoskeleton consists of three structures: microfilaments (MF), microtubules (MT), and intermediate filaments (IF) (Fig. 1.14). The structure, proportion, and distribution of these elements vary among different cells and cell cycles (Fig. 1.15) and are the basis for maintaining cell shape and movement, playing a crucial role in cell mitosis.

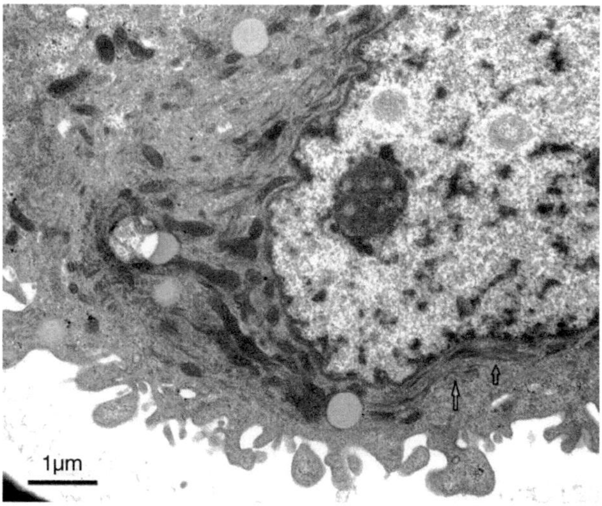

Fig. 1.15 Multiple and layered cytoskeletal structures in squamous carcinoma cells (hollow arrows; TM, 20 000×)

1.2.5 Nucleus and Chromatin

The nucleus is the largest and most important organelle in eukaryotic cells, serving as the regulatory center for cellular heredity and metabolism, and is one of the distinguishing features between eukaryotic and prokaryotic cells (Fig. 1.16). Chromatin is the core component of the nucleus, serving as the carrier of genetic information storage and expression. The nucleus is mainly composed of the nuclear membrane, nuclear lamina, chromatin, nucleolus, and nuclear bodies. The nuclear membrane is the barrier between the nucleus and other organelles and is the only channel for material exchange. It is composed of a double lipid membrane connected to the endoplasmic reticulum, nuclear pores, and the nuclear lamina located on the inner side of the double lipid membrane. The nuclear lamina not only supports the outer nuclear membrane but also has direct contact with the chromatin inside the nucleus (especially heterochromatin). Chromatin is composed of DNA and histones and is classified into euchromatin and heterochromatin based on its degree of condensation, with their distribution related to cell type and developmental state. Chromosomes are a special manifestation of chromatin, which, after replication, form chromosomes that are evenly distributed to two daughter cells. The genes responsible for encoding rRNA are transcribed in the nucleolus, the site of rRNA synthesis and processing, and ribosome assembly. The degree of chromatin condensation varies with cell development, with more euchromatin, more prominent nucleoli, and less heterochromatin in immature cells with active synthesis. Under an LM, these cells

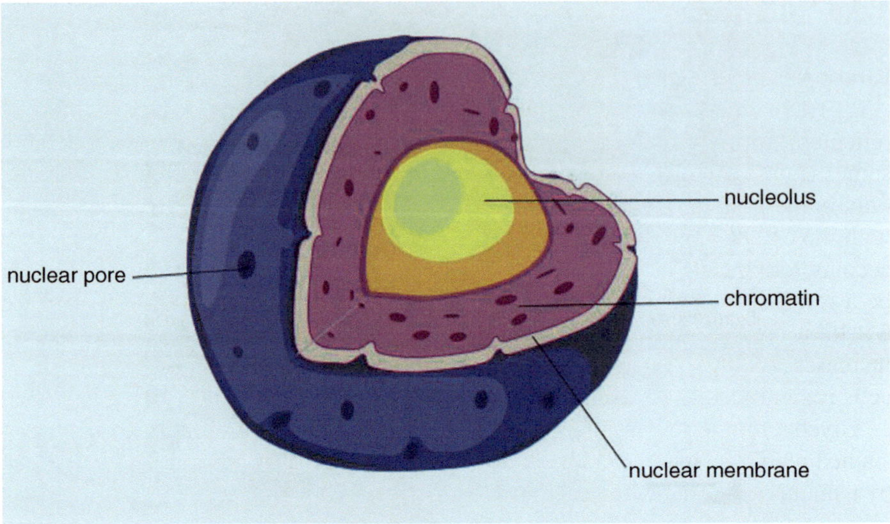

Fig. 1.16 Schematic diagram of the nucleus
核仁——nucleolus；染色质——chromatin；核膜——nuclear membrane；核孔——nuclear pore

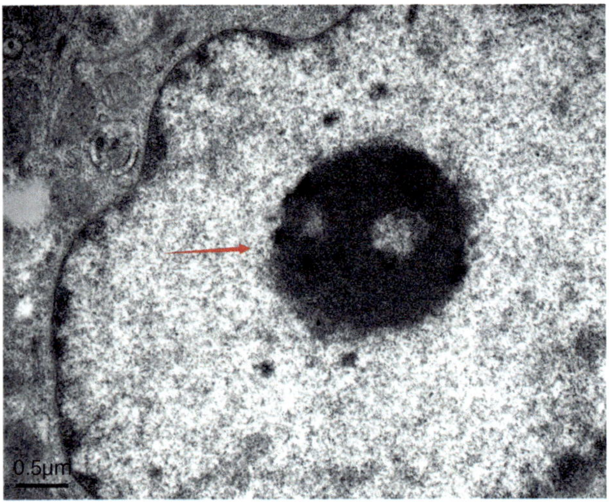

Fig. 1.17 Large bull's-eye-like nuclei of tumor cells of poorly differentiated carcinomas (TM, 30,000×)The nucleus is almost entirely euchromatin, suggesting very active proliferation and differentiation

appear loose and fine, with fewer aggregates and more prominent nucleoli, but the staining conditions can interfere with clear identification. However, under EM, euchromatin appears light-colored, heterochromatin appears as black masses, and the nucleolus appears as a black circular or irregular shape, with less interference from staining, making it easier to distinguish (Fig. 1.17).

1.2.6 Ribosomes

Ribosomes are irregular granular structures mainly composed of RNA and proteins (Fig. 1.18). Many ribosomes attach to the membrane surface of the endoplasmic reticulum, forming rough ER, while some ribosomes are free, known as free ribosomes (Fig. 1.19). They are structurally and chemically identical, with differences only in the proteins synthesized. The number of ribosomes is positively correlated with the extent of protein synthesis; when the number of ribosomes is high, it can account for up to 40% of the cell's weight. In starving cultured cells, there may only be a few hundred ribosomes, while in cultured HeLa cells, the number can reach 5×10^6 to 5×10^7. After staining, the cell color deepens as the number of ribosomes increases. As an indicator of active synthesis, the number of ribosomes in tumor cells is significantly increased.

Glycogen is the form in which glucose is stored in cells. It appears red when stained with PAS and has high electron density under EM, without being enclosed by a membrane. There are usually two types:

- β granules: irregular in shape, with a diameter of 20–30 nm, and dispersed.
- α granules: clusters of β granules, varying in size. In aging leukocytes with low metabolic activity, glycogen granules tend to accumulate due to the inability to be consumed. However, tumor cells with active metabolism and differentiation synthesize large amounts of glycogen, often appearing in clusters (Fig. 1.20), or scattered. Tumor cells with a high degree of differentiation produce more

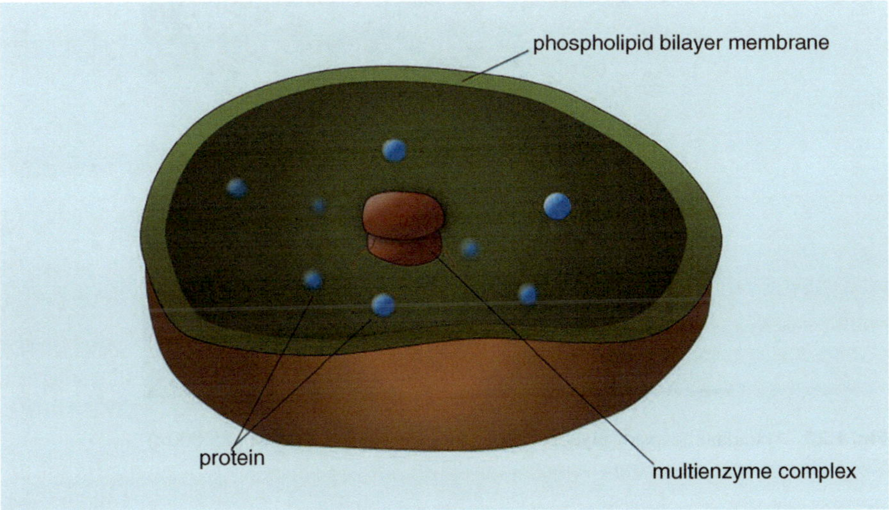

Fig. 1.18 Schematic diagram of a ribosome
磷脂双层膜:phospholipid bilayer membrane； 酶复合物:multienzyme complex； 蛋白质:protein

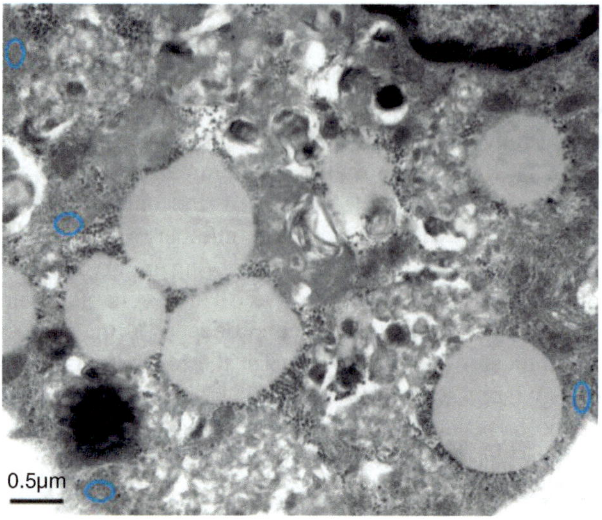

Fig. 1.19 Free ribosomes (TM, 25,000×)
The number of free ribosomes determines the depth of the cell color, usually gray-black in the cell, with particles that are small and inconspicuous, which is the base color of the cell (as marked by the blue circle in the figure), to be distinguished from the clusters or scattered dark black coarser particles of glycogen Glycogen Granules

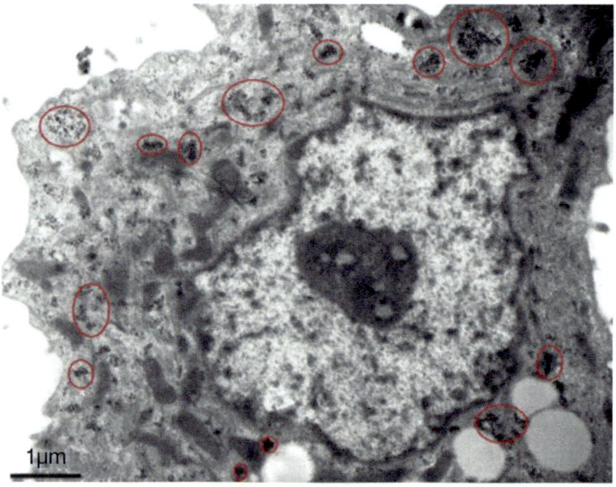

Fig. 1.20 Abundant heaps of glycogen in tumor cells (red circle; TM , 25 000×)

enzymes for synthesis, resulting in more glycogen synthesis and significant accumulation, whereas low-differentiation or undifferentiated tumor cells has insufficient enzyme production, resulting in less glycogen, which is less likely to appear in clusters.

1.3 Morphological Characteristic of Humoral Cells Under Light Microscope

Light microscopy is the foundation for cell structure research and the most widely used tool in various morphological fields. Under a light microscope, the basic structure and morphology of cells can be observed. The study of cell morphology in body fluids started later than in blood, and the morphology and structure are more susceptible to various factors, making them more complex and difficult to identify. Here, we list several common morphological features of cells in body fluids to help understand the differences between light and electron microscopic structures.

1.3.1 Mesothelial Cell

Mesothelial cells are single-layered flat epithelial cells covering the serous membrane surface (Fig. 1.21). In inflammation or other pathological conditions, they can detach into the effusion, often forming inclusion bodies through phagocytosis. Their nuclei are round or oval, with visible nucleoli (usually one), and cytoplasmic volume varies, appearing light blue or blue. Pseudopodia may be present. The most

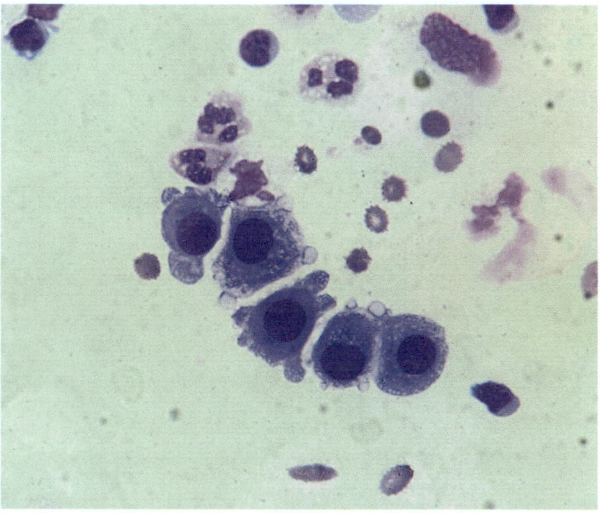

Fig. 1.21 Scattered mesothelial cells (Ritter-Giemsa stain, 1000×)

noticeable difference from macrophages is the round nucleus, more densely packed chromatin, and coarse granules, often with visible nucleoli. Clusters of mesothelial cells form mesothelial pores (Fig. 1.22) with pink homogeneous material in the center, possibly protein secretions from mesothelial cells. The identification of heterogeneous mesothelial cells is challenging under light microscopy and often cannot be clearly distinguished from tumor cells, but it is relatively easy to distinguish under electron microscopy, as described in detail later.

1.3.2 Macrophages

Macrophages (Fig. 1.23) are commonly found in various serous effusions, such as chronic nonspecific inflammation, recovery phase of acute inflammation, tumors, viral, and parasitic infections. They are adept at cleaning up the "battlefield." Macrophages originate from monocytes and possess the characteristics of monocytes. However, macrophages in the body fluid have more irregular cytoplasm compared to monocytes. They are capable of phagocytosing various substances, and their cytoplasm contains noticeable granules, while the nucleoli are hardly visible. In lung tissue, macrophages are referred to as dust cells (Fig. 1.24) because they engulf a large amount of dust particles that enter the lungs. In liver tissue, they are called Kupffer cells. In brain tissue, they are known as microglial cells, which are

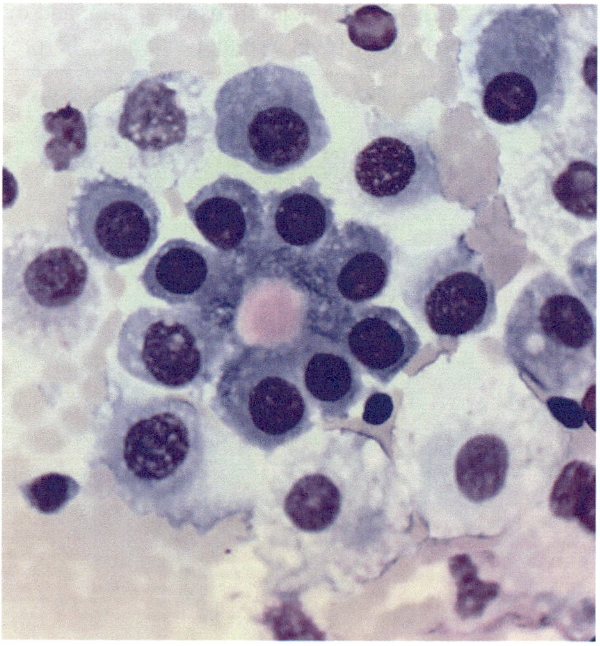

Fig. 1.22 Mesothelial pores formed by mesothelial cells (Ritter-Giemsa stain, 1000×)

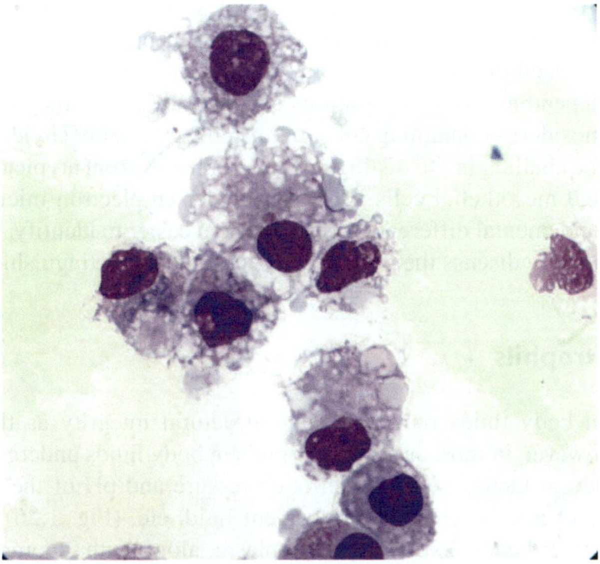

Fig. 1.23 Macrophages (Ritter-Giemsa stain, 1000×)

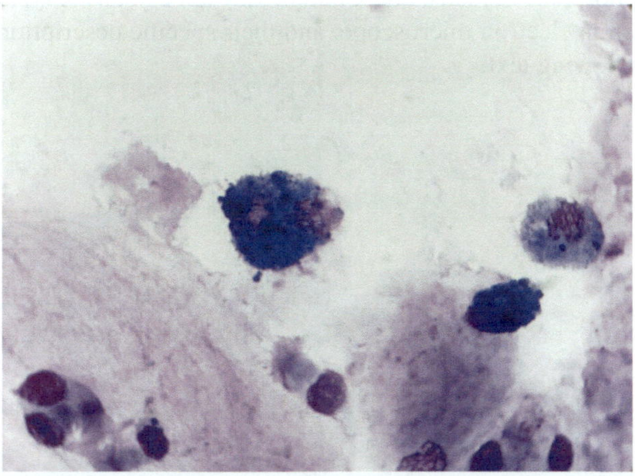

Fig. 1.24 Macrophages in bronchoalveolar lavage fluid, dust cells (Ritter-Giemsa stain, 1000×)

essential cells in the nervous system, playing roles in metabolism, support, protection, nutrition, immunity, repair, and regeneration. The number of microglial cells is ten times that of neurons, and they come in different types. The names of macrophages vary depending on their locations and the substances they engulf, such as dust cells, hemosiderin-containing cells, foam cells, and so on. Under a light microscope, it may be challenging to distinguish macrophages from atypical lymphocytes and degenerated mesothelial cells. However, under an electron microscope, these cells exhibit fundamental differences, making them easier to identify. The following text will specifically discuss their characteristics and key distinguishing features.

1.3.3 Neutrophils

Neutrophils in body fluids have the same structural integrity as those in blood (Fig. 1.25). However, in most cases, neutrophils in body fluids undergo morphological changes due to factors such as osmotic pressure and pH of the fluid, soaking time, presence of microorganisms in purulent fluid, etc. (Fig. 1.26). Pink neutrophilic granules can be observed in the cytoplasm, along with incomplete cytoplasmic structure, appearing as naked nuclei, which distinguishes them from other cells. Various pathogenic microorganisms and crystals can typically be found in the effusion, such as orange-red blood crystals, cholesterol crystals, sodium urate crystals, calcium pyrophosphate crystals, and Charcot-Leyden crystals. Crystals can also be observed under an electron microscope, and their specific descriptions will be provided in the following text.

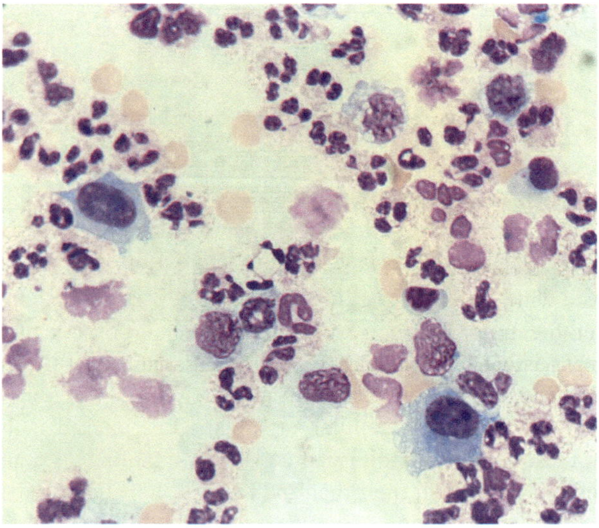

Fig. 1.25 Neutrophils (pleural effusion, Ritter-Giemsa stain, 1000×)

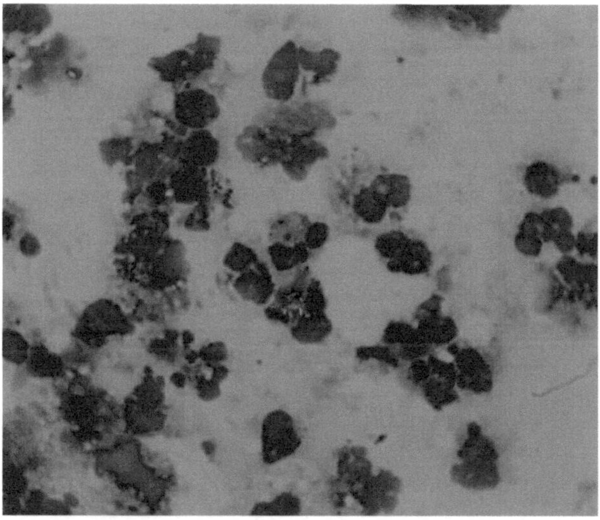

Fig. 1.26 Broken neutrophils in pus (Ritter-Giemsa stain, 1000×)

1.3.4 Eosinophils and Basophils

Eosinophils and basophils play important roles in allergic reactions. Factors such as tumors, tuberculosis, allergies, pneumothorax, and hemothorax are caused by puncture, and parasitic infections can lead to an increase in eosinophils, with or without an increase in basophils.

Eosinophils are highly fragile. When eosinophils increase in number, the broken eosinophilic cells fuse together, forming Charcot-Leyden crystals (Fig. 1.27). Basophils can release histamine and other cytokines, which can induce an increase in eosinophils. Theoretically, basophils that reach the site of action (Fig. 1.28) release cytokines to induce more eosinophils to migrate to the reaction site, resulting in a cascading amplification effect.

1.3.5 Lymphocytes

Lymphocytes are commonly observed in the morphological analysis of serous effusions. Under normal conditions, small lymphocytes in a quiescent state predominate. However, in diseases such as viral or *Mycobacterium tuberculosis* infection, tumors, or systemic lupus erythematosus, an increase in lymphocytes can be seen (Fig. 1.29). With intense stimuli such as viruses or tuberculin, some lymphocytes become activated, leading to an increase in reactive lymphocytes or the appearance of immunoblasts (Fig. 1.30). When a large number of abnormal lymphocytes or immunoblasts are present, it often suggests the involvement of lymphoma in the serous cavity. In the effusion of tuberculosis infection, a significant amount of

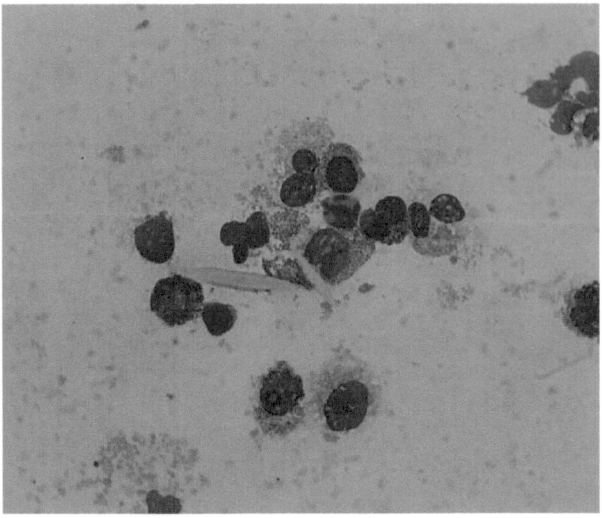

Fig. 1.27 Eosinophils and Charcot-Leyden Crystals (Ritter-Giemsa Stain, 1000×)

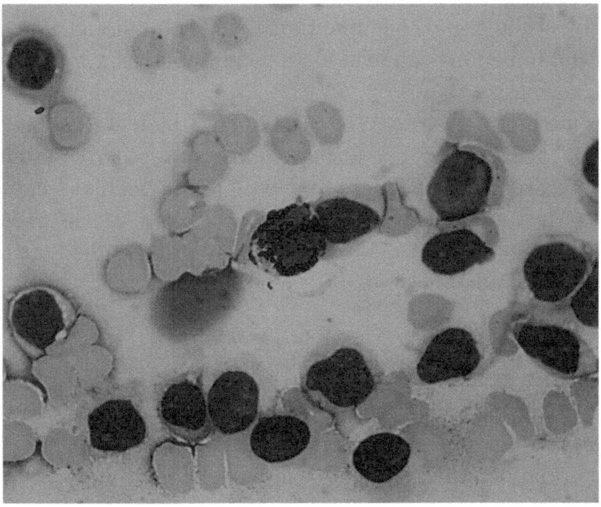

Fig. 1.28 Basophils (Ritter-Giemsa stain, 1000×)

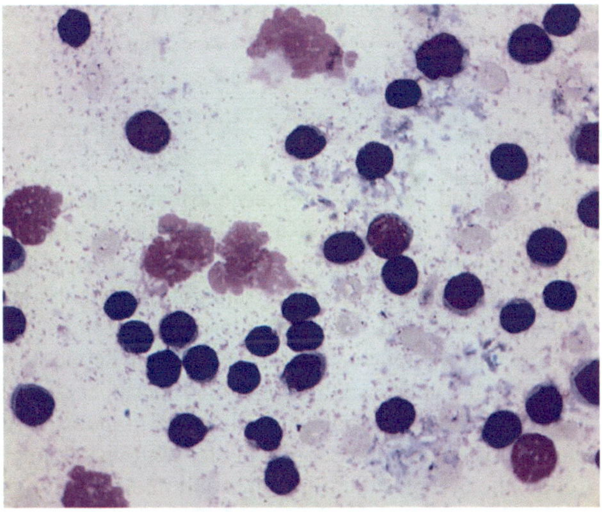

Fig. 1.29 Increased lymphocytes in a tuberculous pleural effusion (Ritter-Giemsa stain, 1000×)

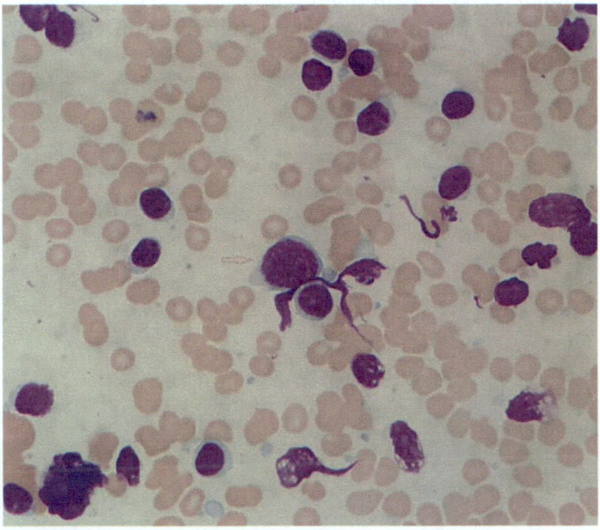

Fig. 1.30 Immunoblasts in lymph node fluid (Ritter-Giemsa stain, 1000×)

necrotic debris may be observed, which can indicate the need for clinical tuberculosis-related investigations.

1.3.6 Plasmatocytes

Plasmatocytes have a round nucleus, often eccentrically located, and a cytoplasm that is relatively abundant. They exhibit a distinct pale-staining perinuclear area under light microscopy, which corresponds to the aggregation of the Golgi apparatus. The well-developed rough endoplasmic reticulum in the cytoplasm is closely related to its strong secretory function. Differentiating plasma cells from lymphocytes and mesothelial cells is not easy under a light microscope. Immunoglobulins secreted by plasma cells appear as pink-stained periphery in the cytoplasm. Lymphocytes have a blue, clear cytoplasm with fewer granules, while mesothelial cells often appear in clusters and have round nuclei with visible nucleoli. Under an electron microscope, the cytoplasmic organelles and nuclear chromatin of different cells are completely distinct, allowing for easy differentiation and providing a good basis for morphological identification.

An increase in plasmatocytes is often classified into two situations: benign reactive proliferation (Figs. 1.31 and 1.32) and malignant proliferation of tumor cells. Benign reactive proliferation usually involves mature plasmatocytes and is characterized by polyclonal increase in immunoglobulins. It is occasionally seen in chronic inflammation or tumor patients. However, malignant proliferation of tumor cells is

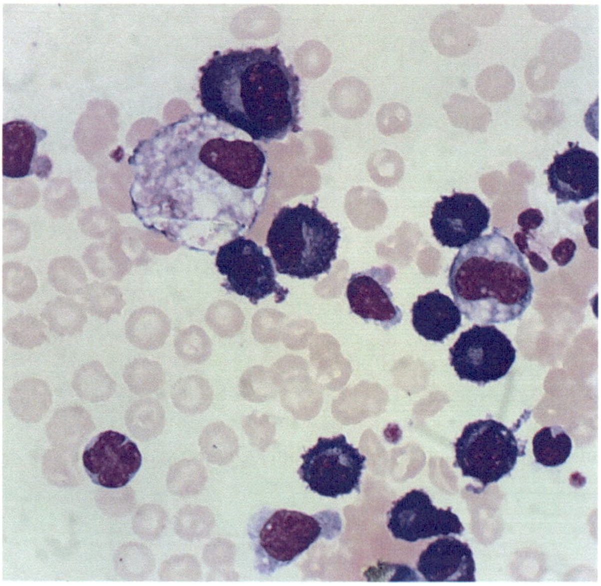

Fig. 1.31 Reactive plasmacytosis (Ritter-Giemsa stain, 1000×)

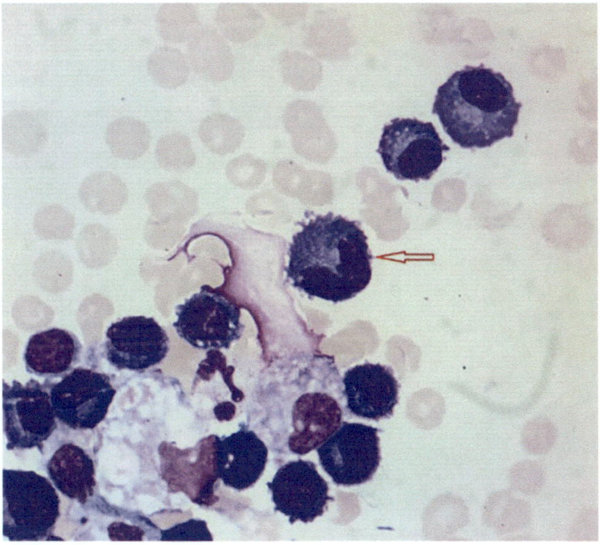

Fig. 1.32 Reactive plasma cells with dikaryotic (Ritter-Giemsa stain, 1000×)

common in myeloma with invasion of the plasma membrane cavity, or in plasma cell leukemia or lymphoplasmacytic lymphoma, which are characterized by increased monoclonal immunoglobulins.

1.3.7 Red Blood Cells

The morphology of red blood cells plays a crucial role in determining the status of bleeding. Freshly hemorrhaged red blood cells appear as biconcave discs with intact morphology (Fig. 1.33), and they are rich in hemoglobin. In specimens from old hemorrhage or with the presence of digestive fluids, red blood cells often appear as crenated cells, fragmented cells, or ghost cells, with incomplete and blurred cell bodies (Fig. 1.34), indicating loss of hemoglobin. Under transmission electron microscopy, red blood cells appear dark due to the dense hemoglobin content, making their morphology easily recognizable (Fig. 1.35). Scanning electron microscopy reveals the complete external morphology of red blood cells (Fig. 1.36).

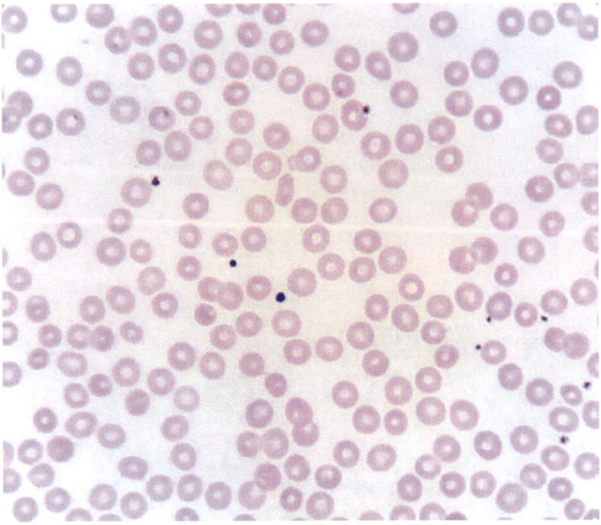

Fig. 1.33 Fresh erythrocytes (Ritter-Giemsa Stain, 1000×)

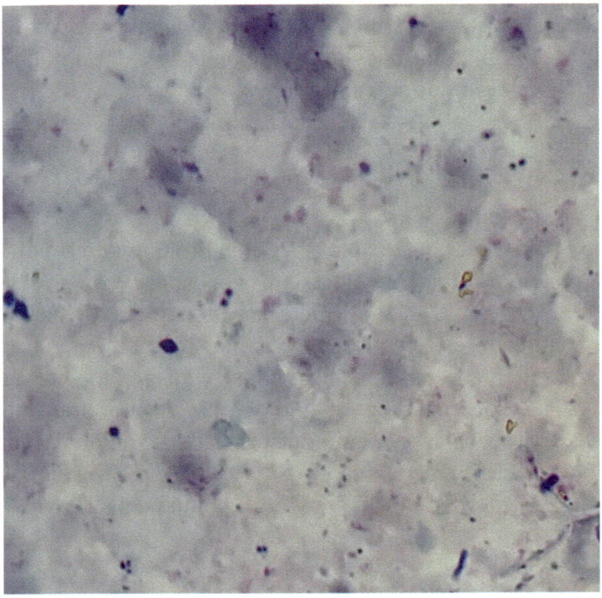

Fig. 1.34 Lysed erythrocytes (Ritter-Giemsa stain, 1000×)

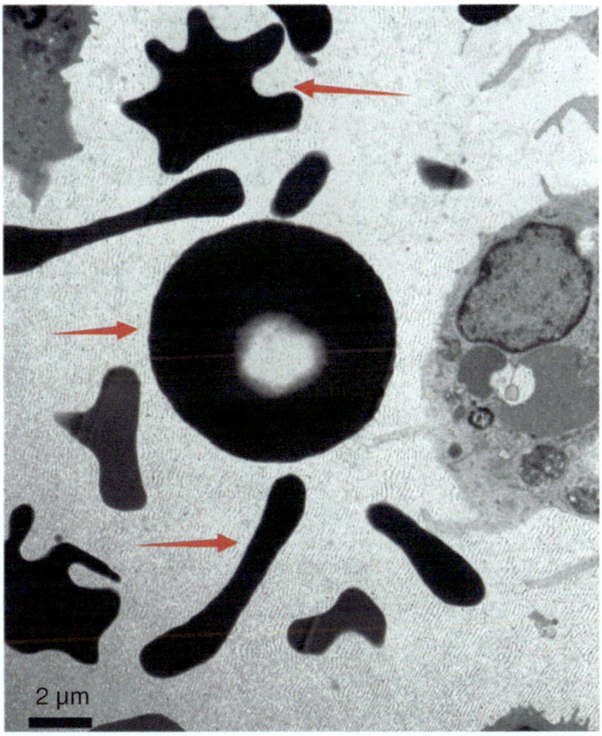

Fig. 1.35 Erythrocytes are black due to dense hemoglobin (TM, 10,000×)

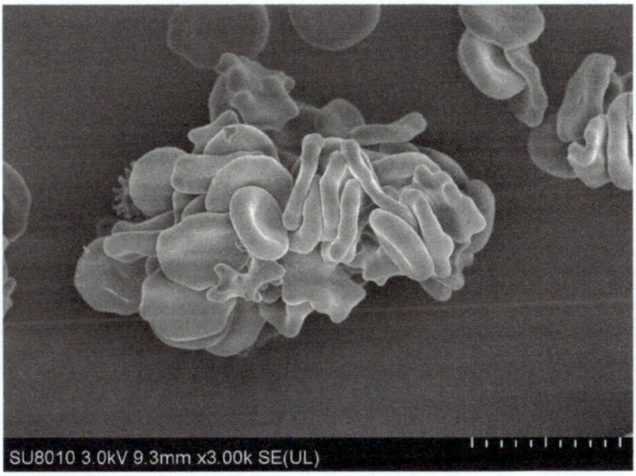

Fig. 1.36 Scanning electron microscopy, erythrocyte clusters, intact erythrocytes in the form of biconcave discs, somatic cytology is prone to spiny erythrocytes and crumpled erythrocytes

2 Comparison of Light and Electron Microscopy of Common Humoral Cells

Dayong Gu and Ying Wang

2.1 Mesothelial Cells

2.1.1 Normal Mesothelial Cells

Under light microscopy (Fig. 2.1), normal mesothelial cells are sometimes difficult to distinguish from macrophages. Under electron microscopy, the nucleus of normal mesothelial cells is round (Fig. 2.2), with a circle of heterochromatin at the edge of the nuclear membrane, and the rest is euchromatin, and the nucleolus is obvious, indicating active cell synthesis; the cytoplasm is irregular, with visible pseudopods and folds, and the cytoplasm is richer in organelles, with visible inclusion bodies and active cytosolic drinking. Macrophages have irregular nuclei, more heterochromatin, no nucleolus or smaller nucleolus, and elongated pseudopods at the cytoplasmic margin, unlike mesothelial cells with blunt rounded pseudopods and folds. Normal mesothelial cells have fewer free ribosomes, lighter cytoplasmic color, and less mitochondria and endoplasmic reticulum, as do nuclear heterogeneous mesothelial cells, thus distinguishing mesothelial cells from tumor cells.

D. Gu (✉) · Y. Wang
Laboratory Department, Shenzhen Second People's Hospital, Shenzhen, Guangdong, China

D. Gu, Y. Wang (eds.), *Morphological Diagnosis of Humoral Cells under Light and Electron Microscopy*, https://doi.org/10.1007/978-981-95-8668-4_2

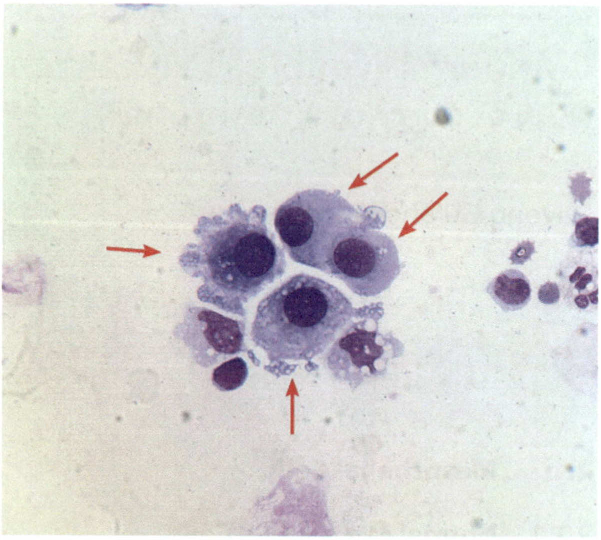

Fig. 2.1 Mesothelial cells (Ritter-Giemsa Stain, 1000×)

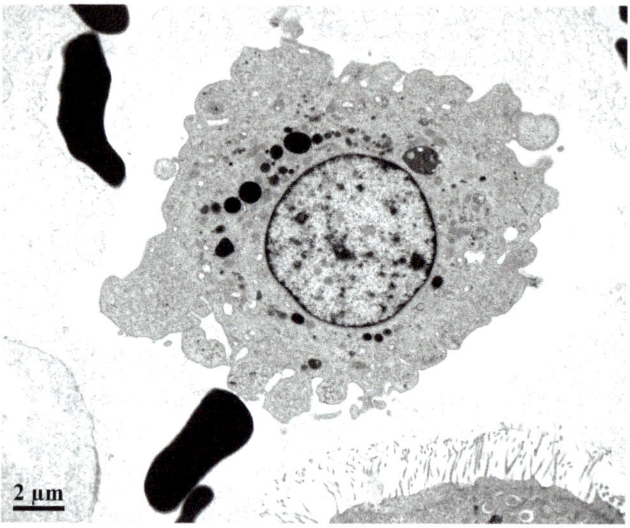

Fig. 2.2 Mesothelial cells (TM, 7000×)

2.1.2 Nuclear Heterogeneous Mesothelial Cells

It is sometimes difficult to distinguish nuclear heteromesothelial cells from macrophages and tumor cells under light microscope (Figs. 2.3 and 2.4). However, under electron microscope (Figs. 2.5), although the nucleus of nuclear heteromesothelial cells is not perfectly round, there is a ring of heterochromatin at the edge of the nuclear membrane with obvious nucleolus. The distribution of chromatin in the

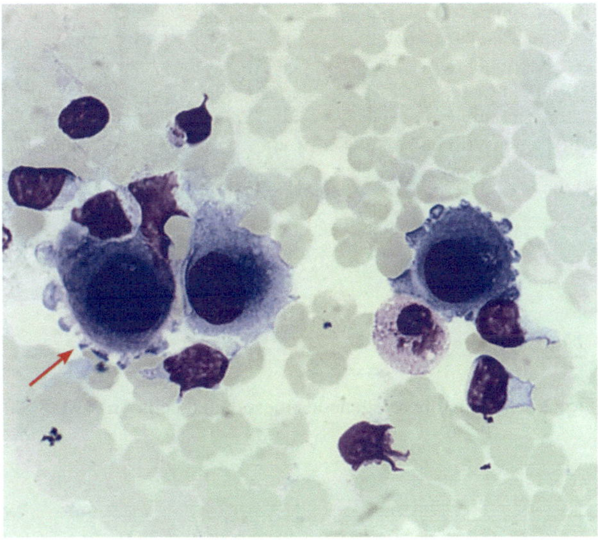

Fig. 2.3 Mild nuclear heterogeneity of mesothelial cells (arrows; Ritter-Gimza stain 1000×)

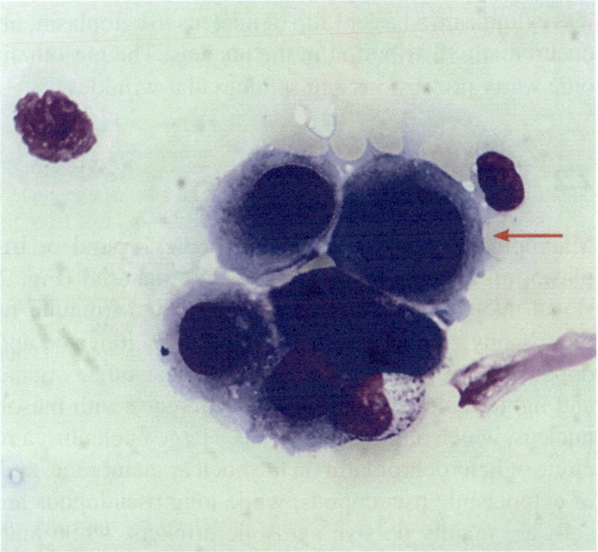

Fig. 2.4 Mesothelial cells with moderate nuclear heterogeneity (arrows; Ritter-Gimza stain 1000×)

nucleus of tumor cells was more uneven. For the tumor cells with low differentiation, the nucleolus were obvious, and the distribution of chromatin was obviously uneven, with euchromatin as the main part, and heterochromatin was less and unevenly distributed around the nuclear membrane. Within the cytoplasm of nuclear heterogeneous mesothelial cells, organelles such as inclusion bodies, mitochondria, free ribosomes, endoplasmic reticulum, and Golgi apparatus were increased, but

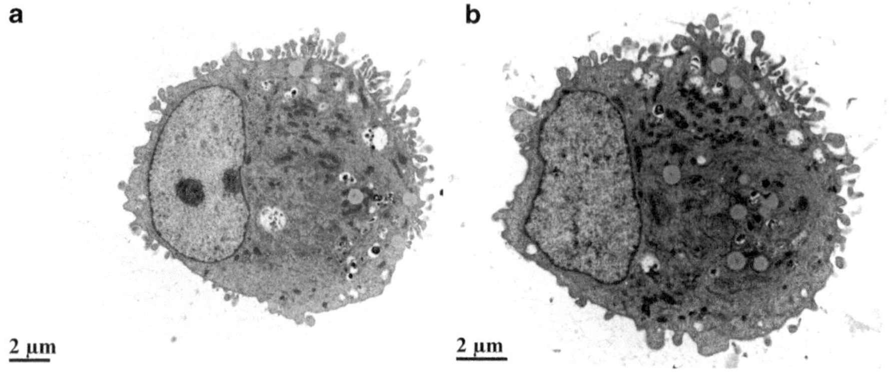

Fig. 2.5 Nuclear heterogeneous mesothelial cells. (**a**) TM, 8000×; (**b**) TM, 10000×

compared with well-differentiated tumor cells, the number was relatively small, and the color was relatively pale. Poorly differentiated tumor cells have less organelles, less cytoplasm, a larger ratio of nucleus to cytoplasm, obvious nucleolus, and mostly euchromatin distribution in the nucleus. The mesothelial cells had typical and specific warty protrusions and semicircular wrinkles.

2.2 Macrophages

Macrophages have irregular cell bodies, round or irregular nucleus, more cytoplasm, more vacuoles and phagocytic material (Fig. 2.6), loose chromatin, occasional nucleolus, mainly euchromatin, and irregular nuclear shape under electron microscopy; activated macrophages have elongated and obvious cytoplasmic pseudopods, more phagocytic vesicles, fewer other organelles, fewer free ribosomes, and lighter color (Fig. 2.7). The difference with mesothelial cells is mainly in the nucleus, which almost always has a large nucleolus, a regular nuclear shape, a thick circle of heterochromatin on the nuclear membrane, and warty protrusions and folds of cytoplasmic pseudopods, while long pseudopods are less common. Mesothelial cells are mainly passive cytosolic drinkers, while macrophages are mainly active phagocytosis, with primary and secondary lysosomes and phagocytic vesicles visible in the cytoplasm.

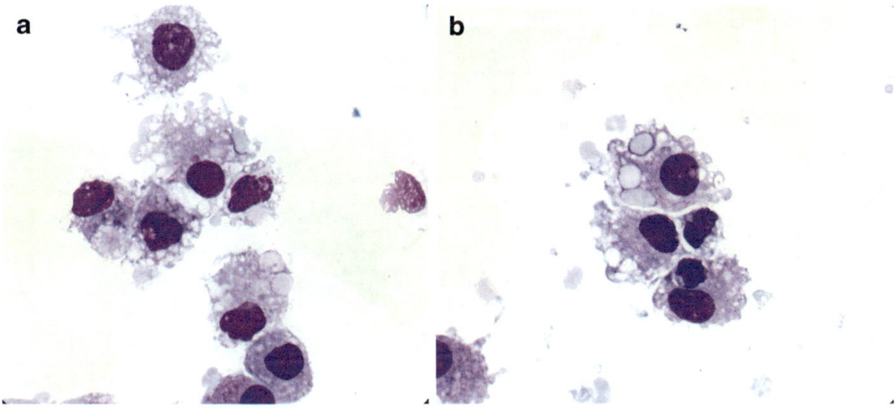

Fig. 2.6 Macrophages (Ritter-Giemsa stain, 1000×)

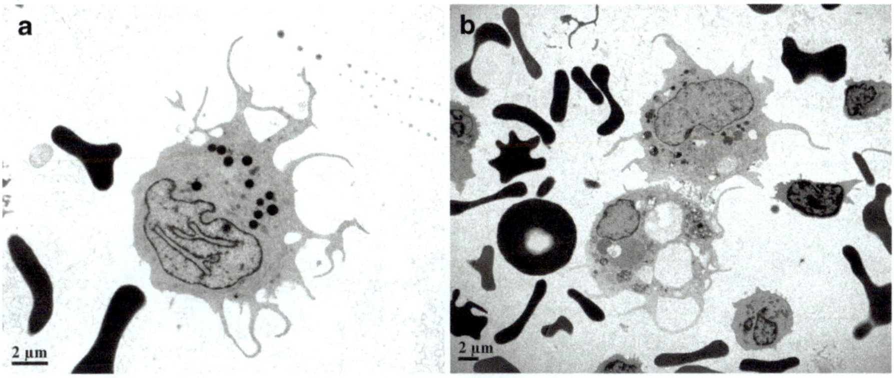

Fig. 2.7 Macrophages (TM, 7000×)

2.3 Neutrophils

The presence of fully intact neutrophils in bodily fluids is rare and typically observed only in new effusions (Figs. 2.8 and 2.9). Under electron microscopy, the neutrophil nuclei are predominantly heterochromatin, aggregated into clumps, with little euchromatin and lobulated nuclei; the cytoplasm has fine enzyme granules, and small vacuoles are visible; the extracellular pseudopods of the cytosol are rare, short, and thick. In prolonged or severe purulent effusions, damaged neutrophils may appear with unclear nuclei, scattered granules, and incomplete membranes when viewed under a light microscope (Fig. 1.26); with electron microscopy, the disconnected membranes, scattered organelles, apoptotic nuclei, and white, sparse chromatin are clearly visible (Fig. 2.10).

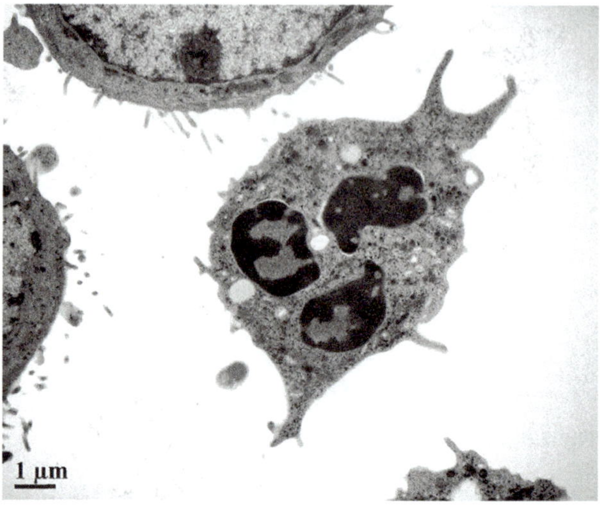

Fig. 2.8 Neutrophils (TM, 12,000×)

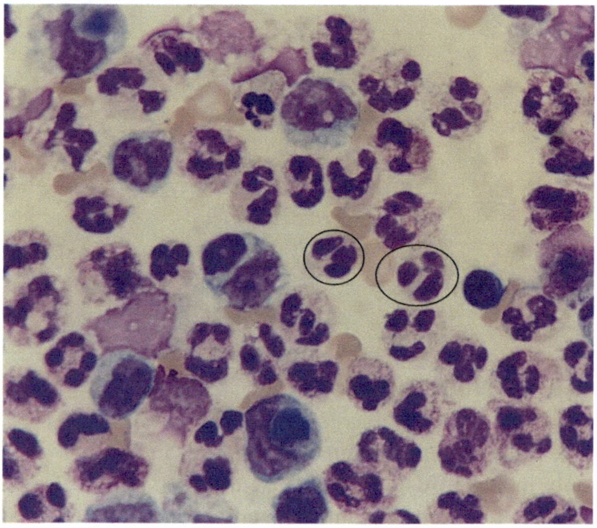

Fig. 2.9 Neutrophils (Ritter-Giemsa stain, 1000×)

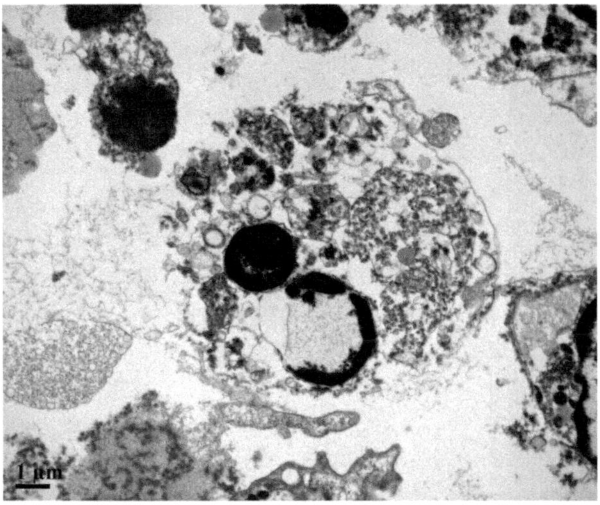

Fig. 2.10 Broken neutrophils (TM, 10000×)

2.4 Lymphocytes

Upon examination of lymphocytes under light microscopy, the appearance of their chromatin is that of a smeary and blocky nature. However, electron microscopy has revealed that this appearance results from large, blocky heterochromatin. Compared to other cells, quiescent lymphocytes have higher levels of heterochromatin and lower levels of euchromatin (Fig. 2.11), which is not apparent under light microscopy but is clearly visible under electron microscopy (Fig. 2.12).

Under electron microscopy, lymphocytes appear more translucent and lighter due to their reduced cytoplasm, organelles, and free ribosomes compared to other cells. The chromatin of activated lymphocytes or lymphoma cells may become naive and loose (Fig. 2.13) and show increased euchromatin and nucleoli under electron microscopy (Fig. 2.14). Identifying the morphology of lymphocytes through light or electron microscopy is insufficient to distinguish between lymphoma cells and naive lymphocytes. Flow cytometry, molecular biology, and immunohistochemistry are required to determine their identity accurately.

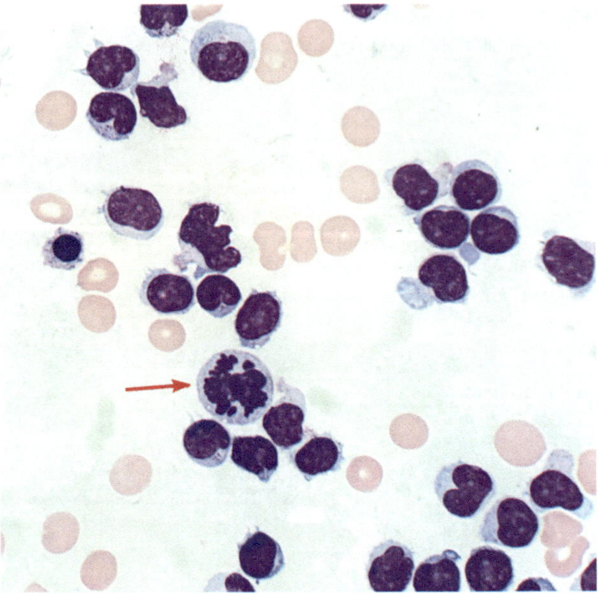

Fig. 2.11 Lymphocytes and their nuclear schizonts (arrow; Ritter-Giemsa stain, 1000×)

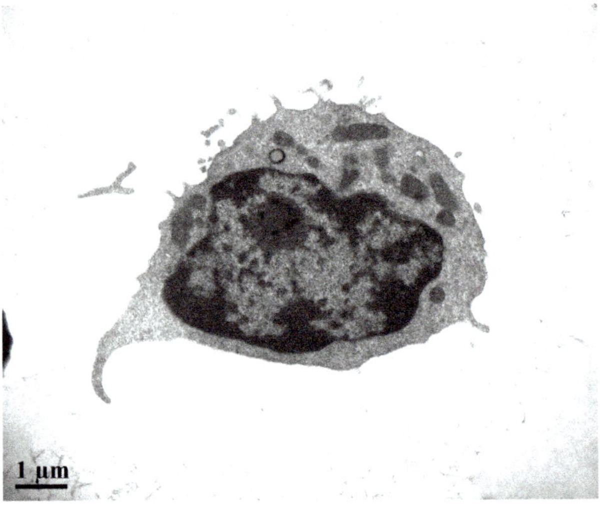

Fig. 2.12 Lymphocytes (TM, 15,000×)

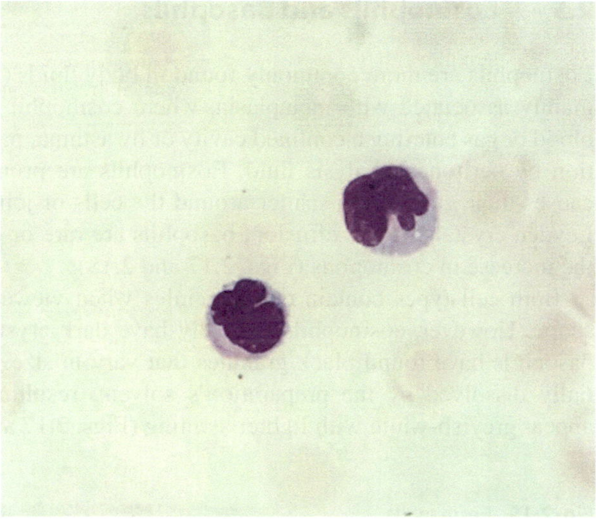

Fig. 2.13 Reactive Lymphocytes (Ritter-Giemsa stain, 1000×)

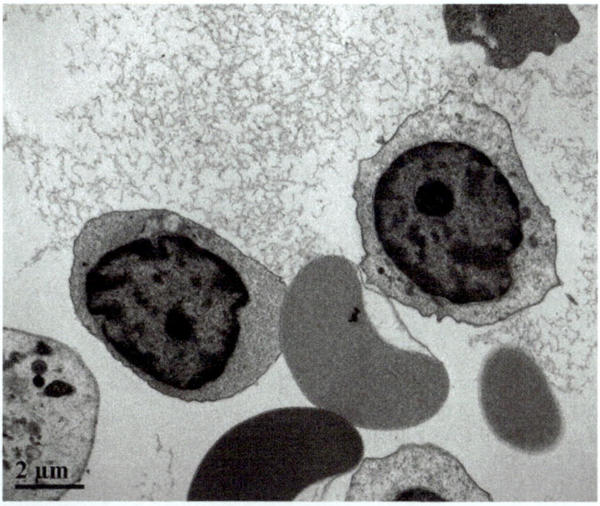

Fig. 2.14 Reactive lymphocytes with lobulated nuclei and distinct nucleoli (TM,10 000×)

2.5 Eosinophils and Basophils

Eosinophils are more commonly found in body fluids (Figs. 2.15 and 2.16) and are mainly associated with metaplasia, where eosinophil chemotaxis is increased by blood or gas entering a confined cavity or by asthma, parasitic infection, or stimulation by peritoneal dialysis fluid. Eosinophils are prone to breaking apart, which causes their granules to scatter around the cells or join together to form Charcot-Leyden crystals. In the effusion, basophils are rare or occasional and often follow the increase in eosinophils (Figs. 2.17 and 2.18).

Both cell types contain dark granules when viewed under an electron microscope. However, eosinophils typically have dark crystals in their granules, while basophils have round black granules that vary in size. These granules can be partially dissolved by the preparation's solvent, resulting in vacuolated areas that appear greyish-white with lighter staining (Figs. 2.17 and 2.18).

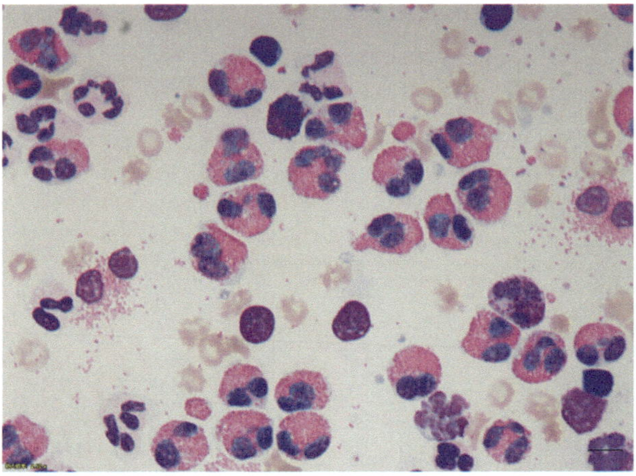

Fig. 2.15 Eosinophils (Ritter-Giemsa stain, 1000×)

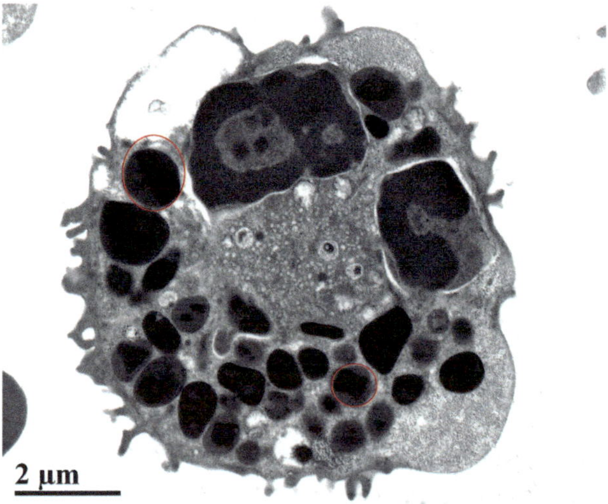

FIG. 2.16 Eosinophils, with crystals in eosinophil granules in the red circle (TM, 15,000×)

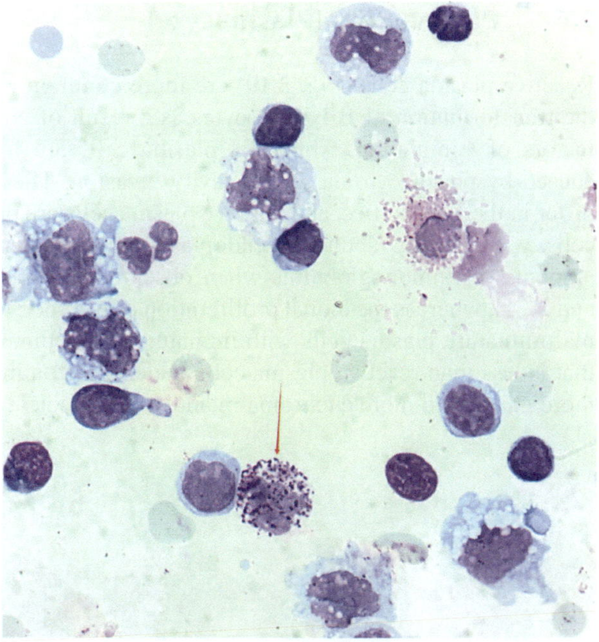

Fig. 2.17 Basophils (Ritter-Giemsa stain, 1000×)

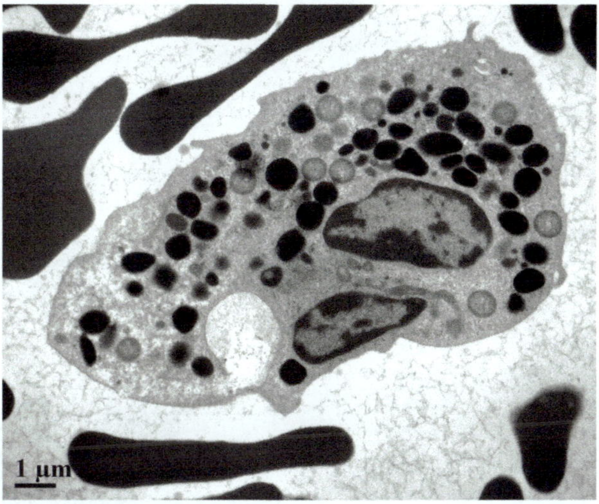

Fig. 2.18 Basophils (TM, 12,000×)

2.6 Plasma Cells (Plasmacyte)

Reactive plasma cells (Fig. 2.19) are more common in body fluids, mainly due to the transformation of B lymphocytes as a result of chronic inflammatory or other factors of stimulation. Abnormal plasma cells are found as clonal proliferation caused by plasmacytoma serous cavity invasion. These two types of plasma cells differ in their properties, and reactive plasma cells are predominantly mature plasma cells with greatly abundant endoplasmic reticulum, which can produce large amounts of immunoglobulins when observed in electron microscopy (Figs. 2.20 and 2.21), whereas the clonal proliferation of plasmocytoma is dominated by abnormal immature plasma cells with immature development of endoplasmic reticulum that is less than reactive plasma cells under electron microscopy, whose nucleus is more naive with more euchromatin and visible nuclei (see cases).

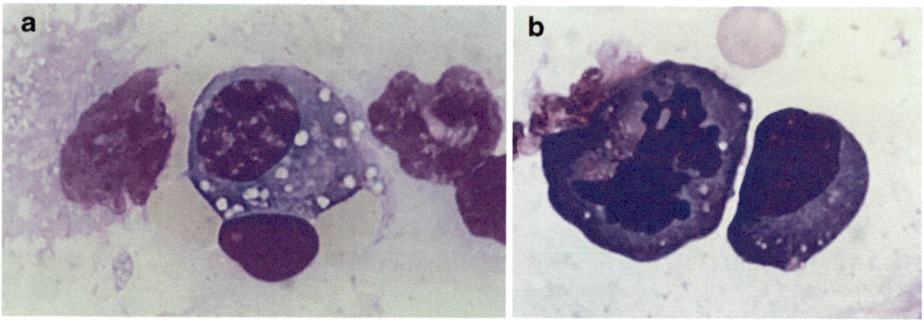

Fig. 2.19 Plasma cells (Ritter-Giemsa Stain, 1000×)

Fig. 2.20 Abundant endoplasmic reticulum in the cytoplasm of mature plasma cells (TM, 15000×)

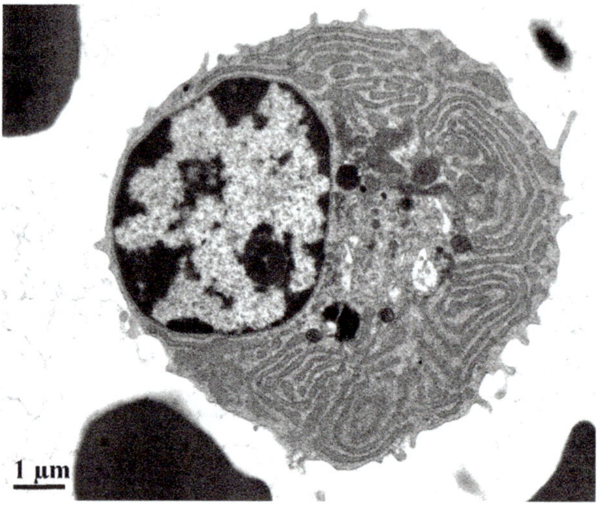

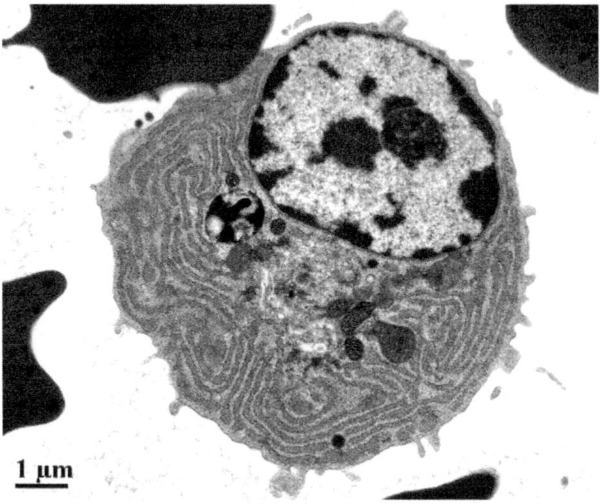

Fig. 2.21 Mature plasma cells with perinuclear Golgi bodies and mitochondrial aggregates (TM, 15000×)

2.7 Ciliated Columnar Epithelial Cells

In alveolar lavage fluid, the percentage of epithelial cells should not exceed 5% in well-taken specimens, while the increase in ciliated columnar epithelial cells may be resulted in the epithelium of the upper superior segmental bronchus brought in when sampled by bronchofiberscope, or the hedding of ciliated columnar epithelial cells in clusters caused by edema and inflammation, which can be described in the report elaborately (Figs. 2.22 and 2.23).

The ciliated columnar epithelial cells are long columnar with nucleus at one end, coarse reticulation of chromatin, no or small nucleoli appeared, and densely clustered cilia at the other end when observed in electron microscopy (Fig. 2.24); the ciliated columnar epithelial cells under electron microscopy were hard to capture ciliated cells and structures that are side-sectioned intactly due to small cilia and different sections, there are fewer free ribosomes in cells and lighter in color (Fig. 2.25).

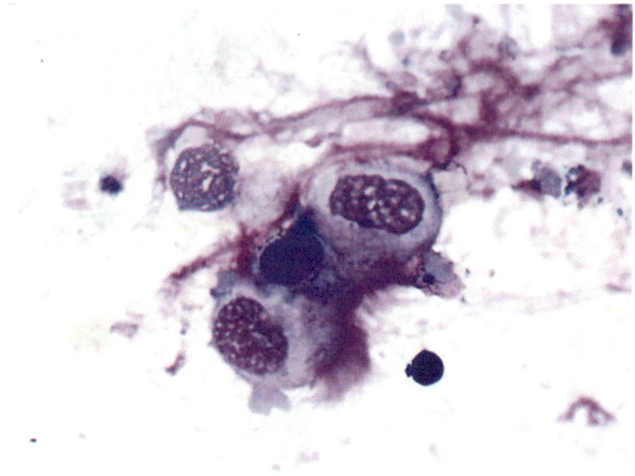

Fig. 2.22 Swollen masses of ciliated columnar epithelial cells (Ritter-Giemsa stain, 1000×)

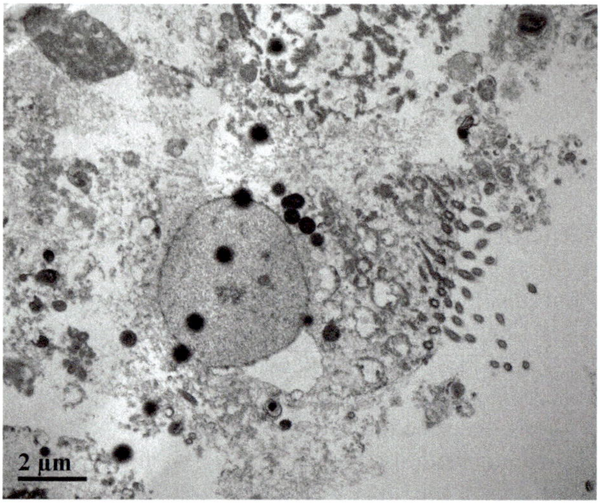

Fig. 2.23 Apoptotic ciliated columnar epithelial cells (TM, 10 000×)

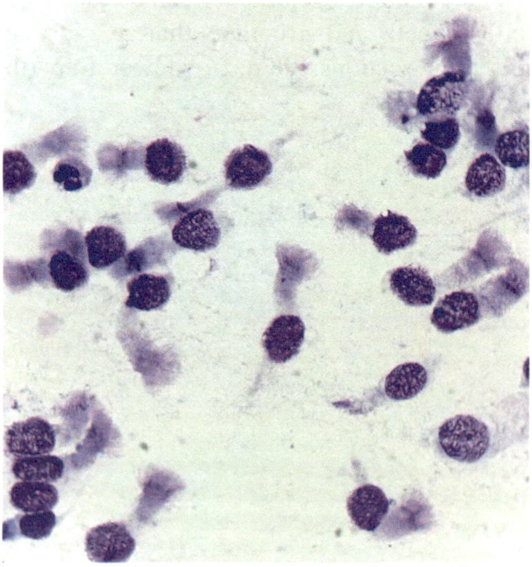

Fig. 2.24 Normal ciliated columnar epithelial cells (Ritter-Giemsa stain, 1000×)

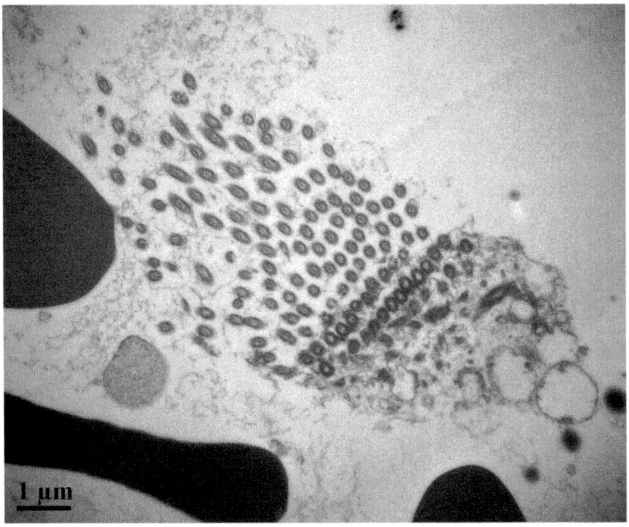

Fig. 2.25 Cilia of ciliated columnar epithelial cells (TM, 15 000×)

2.8 Iron-Containing Hematoxylin Phagocytes and Dust Cells

Both iron-containing hematoxylin phagocytes and dust cells are macrophages, named for their diverse phagocytic contents. Iron-containing hematoxylin phagocytes (Fig. 2.26) are macrophage who phagocytoses red blood cells to form ferritin-containing particles suggesting old hemorrhage; dust cells (Fig. 2.27) are formed by macrophages swallowing dust or powder and other substances that cannot be decomposed, which are more common in smokers and patients with occupational pneumoconiosis. Smokers can see black particles swallowed by dust cells and blue lipid droplets formed by tar. The number of particles correlates with smoking history, dust cells, and iron-containing hematoxylin phagocytes may be examined in alveolar lavage fluid, and iron-containing hematoxylin phagocytes can also be found in other old hemorrhagic effusions. The granules of both types of cells were black, the black of iron-containing hematoxylin phagocytes was different in heterogeneity and size (Fig. 2.28), and the dust cell granules are observed in gray, black, or other shades of color and varied in size. Iron-containing hematoxylin phagocyte granules are blue, and dust cells are black-gray when stained with Prussian Blue Iron, and the difference was not obvious when other staining was performed (Fig. 2.29).

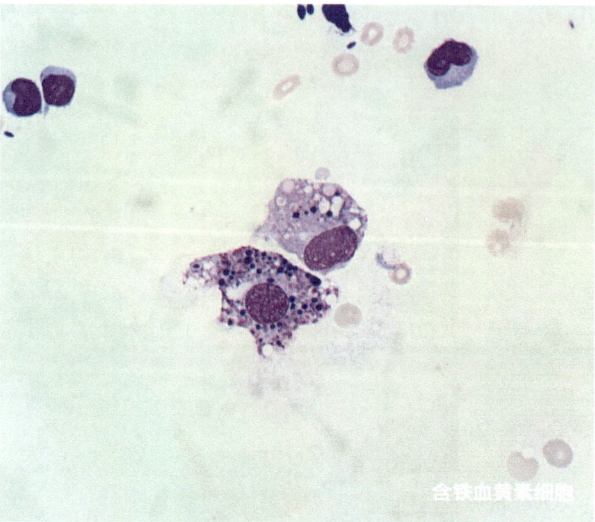

Fig. 2.26 Iron-containing hematoxylin phagocytes (Ritter-Gimza staining, 1000×)

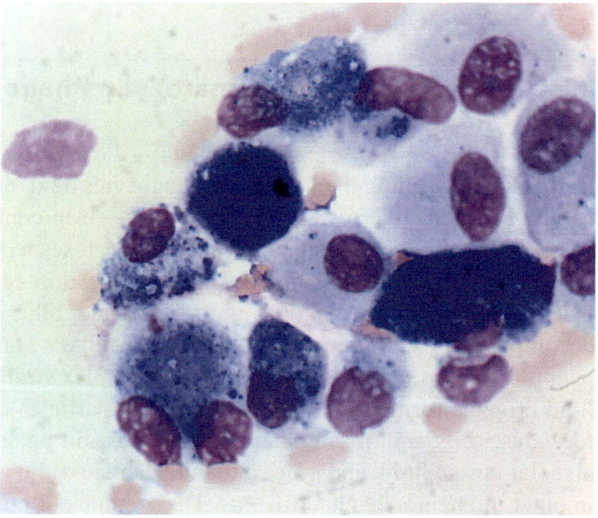

Fig. 2.27 Dust cells (Ritter-Giemsa Stain, 1000×)

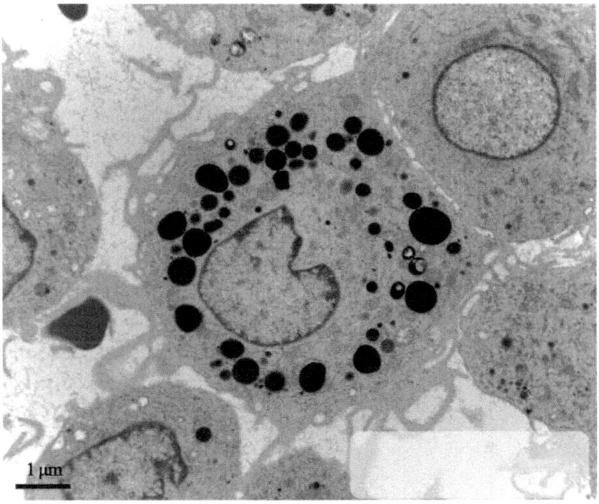

Fig. 2.28 Iron-containing hematoxylin phagocytes (TM，8000×)

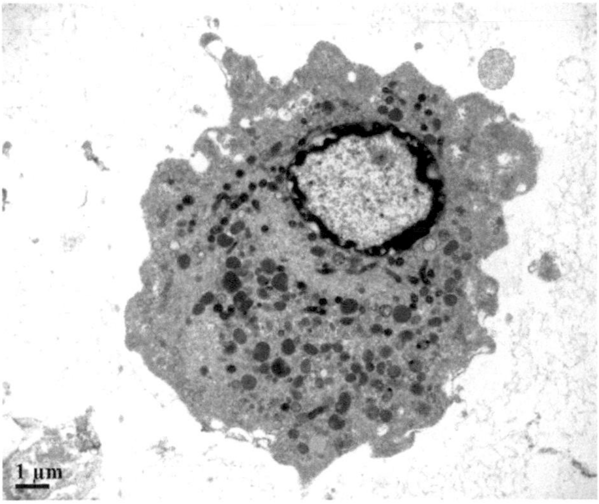

Fig. 2.29 Dust cells (TM，10 000×)

Light and Electron Microscopy Comparison of Adenocarcinoma

3

Dayong Gu and Ying Wang

The classification of tumor cells follows the classification principle of pathology. The diagnosis of tumors requires a variety of means of comprehensive judgment; morphology cannot ultimately distinguish the category of tumors, but it can clarify the existence of tumors or suggest a high degree of clinical suspicion; and it is necessary to carry out a series of examinations, such as imaging and pathology, and to seek evidence from various sources in order to confirm the diagnosis. The value of body fluid cell morphology lies in early diagnosis and rapid diagnosis, which cannot be replaced by other tests.

The same tumor, the same genotype, because the internal environment is different, the morphology of the tumor cells is also very different, but the tumor cell characteristics have their own rules to follow, such as light microscopy can be manifested as cell deep staining; cell body is huge; the nucleo-cytoplasmic ratio is large; the nucleolus is obvious; the nuclear aberration is obvious; the cytoplasm is cloudy; the vacuoles are many, uneven, phagocytosis phenomenon; the cell arrangement appears as a group; and cell plasma boundaries are not clear. However, some reactive nuclear heterogeneous cells are difficult to distinguish from tumor cells.

Electron microscopy is commonly used in foreign laboratories to assist diagnosis, and electron microscopy can more clearly show the ultrastructure of cells, for example, the proliferation of tumor cells is manifested in the number of free ribosomes. It has been reported that the free ribosomes starved in vitro cultured cell could be as few as a hundred, while Hela cells could have $5 \times 10^6 \sim 5 \times 10^7$. Due to the high protein content of free ribosomes, it is strongly alkaline (blue) when stained with Ritter-Giemsa stain and black when stained with heavy metal, so the tumor cells often show dark blue cytoplasm when observed by light microscope, while the color of proliferating tumor cells is dark gray when observed by electron microscope, and the free ribosomes are scattered all over the cytoplasm, the particles are

D. Gu (✉) · Y. Wang
Laboratory Department, Shenzhen Second People's Hospital, Shenzhen, Guangdong, China

D. Gu, Y. Wang (eds.), *Morphological Diagnosis of Humoral Cells under Light and Electron Microscopy*, https://doi.org/10.1007/978-981-95-8668-4_3

fine and black, and the low-differentiated tumor cells have relatively fewer free ribosomes, so they show a lighter color. In addition, there is the number of mitochondria, which provide ATP that is essential for cellular activity and necessary for cell proliferation or growth, and a higher number of mitochondria indicates greater activity, proliferation, and growth. Tumor cells usually have many more mitochondria than normal cells, which can only be seen by electron microscopy. The number of endoplasmic reticulum and Golgi apparatus is also one of the important aids. The endoplasmic reticulum is the main factory for the production of proteins and lipids, and the Golgi apparatus is the site of further processing and synthesis, and its number is also closely related to the cell's secretion-synthesis function, and the endoplasmic reticulum is markedly increased in the plasma cells, which is also a reflection of its function. The amount of glycogen is an indicator to distinguish benign and malignant. In malignant tumor cells, the synthesis and reserve of glycogen are increased, and you can see clusters of aggregated α-type glycogen and scattered and densely distributed β-type glycogen, and under the electron microscope, the glycogen is in the form of black particles, and the increase in its amount makes the cells darker in color. The cytoskeleton and bundles of tonofibrils can be used as a differential indicator of squamous carcinoma, and a marked increase in the number of tonofibrils, appearing in bundles, can be clearly observed in the more differentiated squamous cell carcinomas. The number of secretory vesicles and lipid droplets is obviously increased, and the nucleus is often pushed to one side in an oblique position, which can be used as a differential indicator of adenocarcinoma. The number and length of microvilli or pseudopods on the cell surface are closely related to the implantation metastasis and invasion ability of tumor cells. Due to the proliferation of tumor cells, the proportion of euchromatin increases, nucleoli are large and clear, chromatin distribution is uneven, and the shape of nuclei is irregular and different.

3.1 Adenocarcinoma

Adenocarcinoma is a malignant tumor derived from glandular epithelium, which may exhibit growth patterns such as acinar, papillary, bronchioloalveolar, or solid. In body fluids, adenocarcinoma cells may originate from in situshedding or represent metastatic spread. The characteristics of these cells are closely associated with the function of their primary tissue and organ. They may present as a uniform population or as a mixture of cells with varying degrees of differentiation or from different origins. Adenocarcinoma cells are often large, with abundant cytoplasm and eccentrically located nuclei. Their acinar structures vary in size and are irregularly arranged. Nuclear pleomorphism is common, with cells showing single, double, or multiple nuclei of varying dimensions. Cell borders are often indistinct, and the cells tend to cluster. Under light microscopy, chromatin appears loose; under electron microscopy, euchromatin predominates, nucleoli are visible, and organelles are abundant, often arranged in clusters. Secretory vesicles differ depending on the tumor's origin. Those arising from secretory organs typically contain numerous small vesicles, whereas tumors from other origins may show larger and irregularly shaped vesicles. The following sections describe several types of adenocarcinoma cells from different primary sites.

3.1.1 Adenocarcinoma of Lung

Case: Patient, 83 years old male, acute course of disease. Two months ago, there was no obvious cause of cough and sputum, which was not easy to cough up, accompanied by chest tightness after activity and got better after resting. No discomfort such as fever, hemoptysis, nausea, and vomiting. Chest CT showed partial consolidation in the left lung and left lung bronchus stenosis, occlusion, more advanced than before. Bilateral pleural effusion, improved and discharged after anti-infective treatment. After discharge, he still had cough and sputum, cough aggravated, obvious chest tightness after activity, with low-grade fever, no dizziness and headache, no lower extremity edema. Rehospitalization, blood routine +CRP showed elevated leukocytes and neutrophils, elevated CRP, tumor markers were significantly elevated, and B-ultrasound suggested a right pleural effusion. Diagnosis considers "lung tumor, obstructive pneumonia." After excluding of contraindications, thoracic puncture and catheterization were performed on May 14, 2021. Routine and biochemical indications of pleural effusion indicated exudation and considered malignant pleural effusions due to tumors first. Morphology of humoral cells, as shown in Figs. 3.1, 3.2, 3.3, 3.4, 3.5, 3.6, 3.7, 3.8, 3.9, and 3.10.

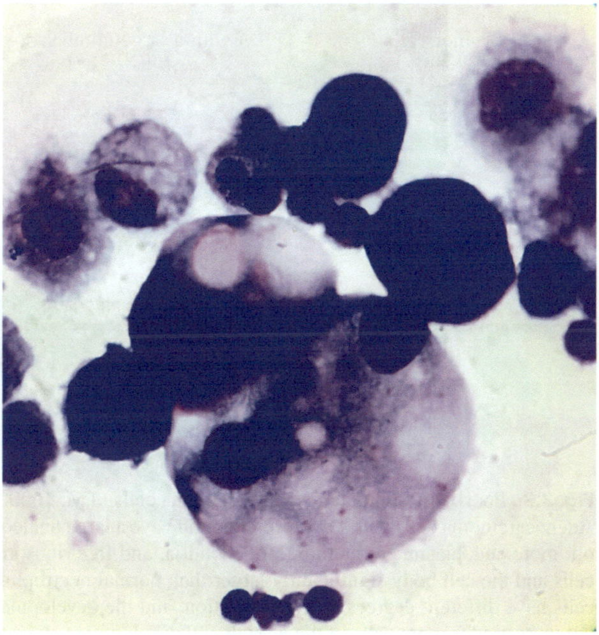

Fig. 3.1 Adenocarcinoma cells (Ritter-Giemsa stain, 1000×)

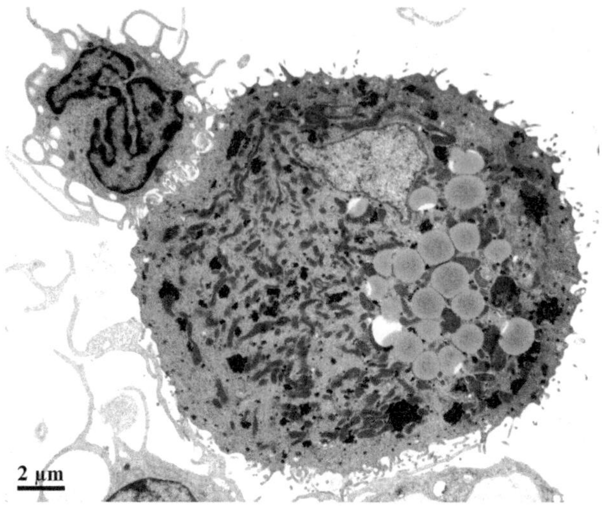

Fig. 3.2 Adenocarcinoma cells and macrophages (TM, 7000×)
Tumor cells and macrophages interact with each other, and the tumor cells have large cell bodies and nuclear malformation. It is incomplete display because of sectional reasons and euchromatin is predominant. A large number of free ribosomes in the cytosol are grayish-black. Glycogen appears in clusters, dense and dark in color. Mitochondria are abundant and vary in shape. Lipid secretion bubbles are relatively small, but the amount is large. There are a large number of short, dense microvilli outside the cell, some of which interacting with macrophages, transmitting information and substances to each other through connecting gaps

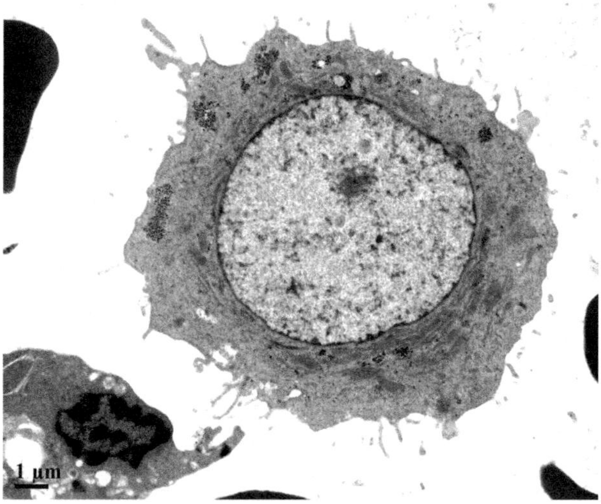

Fig. 3.3 Poorly differentiated adenocarcinoma cells (TM, 10000×)
Adenocarcinoma cells with low differentiation have a large nucleocytoplasmic ratio, obvious nucleoli, more endoplasmic reticulum, mitochondria, and free ribosomes in the cytoplasm than normal cells and the cell body significantly larger than normal neutrophils in the lower left corner. Tumor cells have different degrees of differentiation, and the development law is similar to normal cells, but there are far more cells in the growth and development stage than normal cells. The renewal of normal cells is regular, polar, and relatively neat, while tumor cells proliferate malignantly, grow and divide disorderly, and arrange disorderly, and the plasma boundary is not clear, often in piles

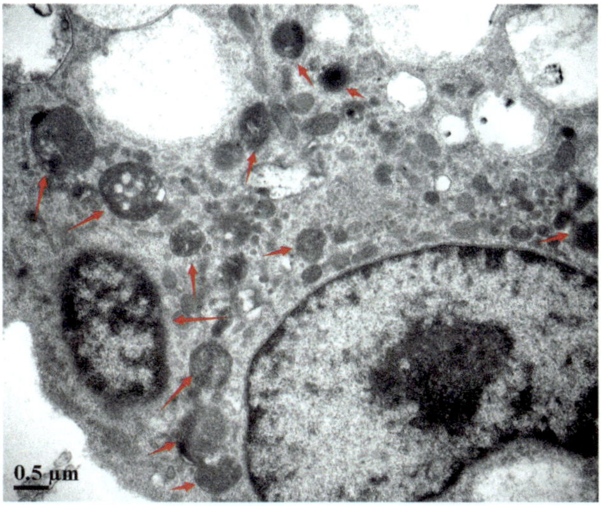

Fig. 3.4 Abundant autophagy or phagocytic vesicles (arrows) in adenocarcinoma cells (TM, 20000×)
Tumor cells' autophagy or phagocytic bubbles (arrows). The autophagy effect plays an important role in maintaining the stability of their own internal environment and strong self-repair. In the process of metastasis, repair damage and remove their own aging and degenerative organelles, maintain their own stability, and maintain vigorous proliferation; tumor cells can quickly engulf, digest, and decompose necrotic, senescent organelles or tissue components and then reassemble them into new organelles, and their repair ability is amazing, thereby maintaining the proliferation and aggressiveness of tumor cells

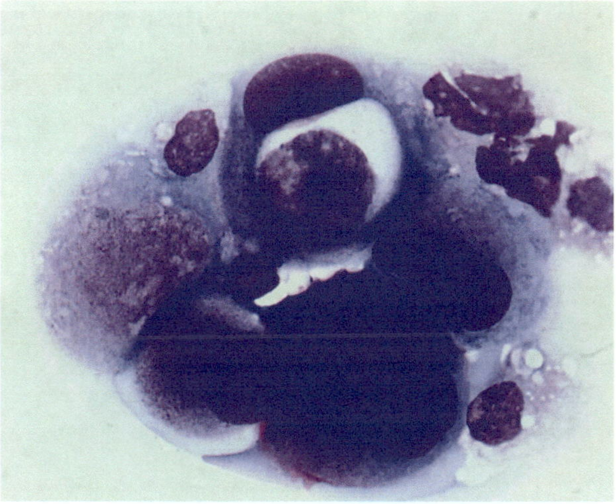

Fig. 3.5 Clusters of adenocarcinoma cells (Ritter-Giemsa staining, 1000×)
Under the light microscope, tumor cells show large cell bodies, deeply stained nuclei, and cloudy cytoplasm. Large inclusion bodies are visible. Tumor cells are piled up into clusters, and the boundaries between cells are unclear. The organelles in the cytoplasm cannot be identified under the light microscope

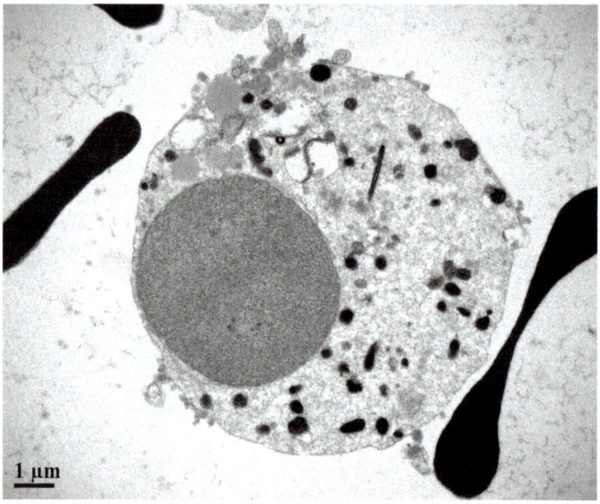

Fig. 3.6 Apoptotic adenocarcinoma cells (TM, 12,000×)
The apoptotic tumor cells have a large inclusion body, obvious acinus in the cytoplasm, and a large number of mitochondria, which belongs to the same group as other tumor cells, and the nucleus is not cut in this plane. In cytology of body fluids, if the cells remain in the fluid for a long period of time, the morphology of the cells is not complete after staining, and the cell type can be distinguished by the residual organelles and structures

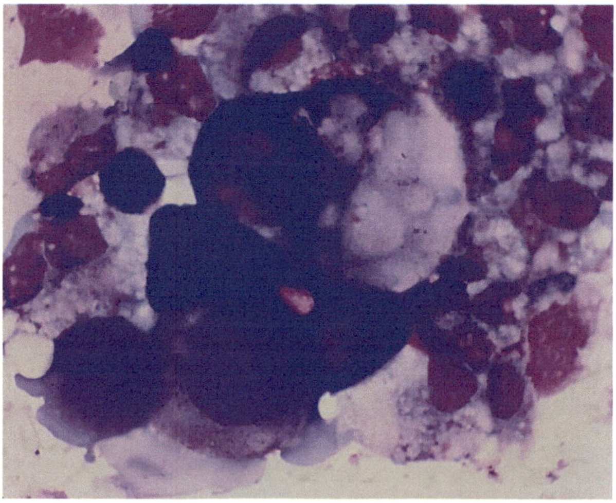

Fig. 3.7 Arranged adenocarcinoma cells (Ritter-Giemsa stain, 1000×)

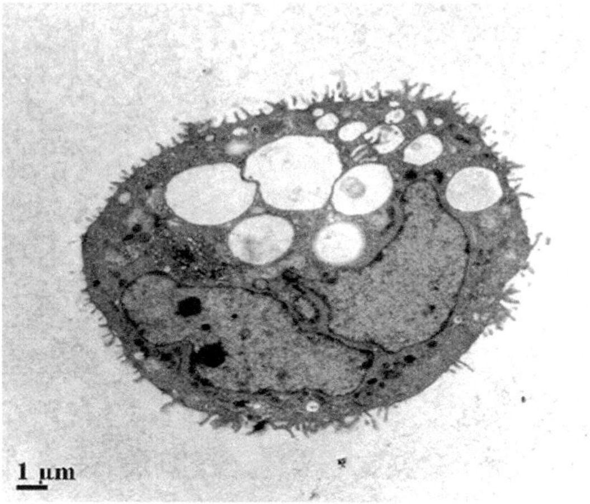

Fig. 3.8 Adenocarcinoma cell (TM, 8000×)
Adenocarcinoma cells have characteristic acinus structures, varying in number and size. Nuclear malformations are pronounced, mainly in euchromatin and one or more nucleoli. Organelles are abundant in the cytoplasm, free ribosomes and glycogen are abundant, and the cells are gray-black. The microvilli outside the cell membrane are rich and fine. Compare with Figure 3.7. Adenocarcinoma cells are larger than surrounding macrophage somas, appearing in piles and being varied in size, among which tumor cells with poorly differentiation levels have unobvious acinus and have a large nucleo-cytoplasmic ratio. The well-differentiated adenocarcinoma cells have small nucleo-cytoplasmic ratio, large cell body, many acinus, and uneven distribution and different size

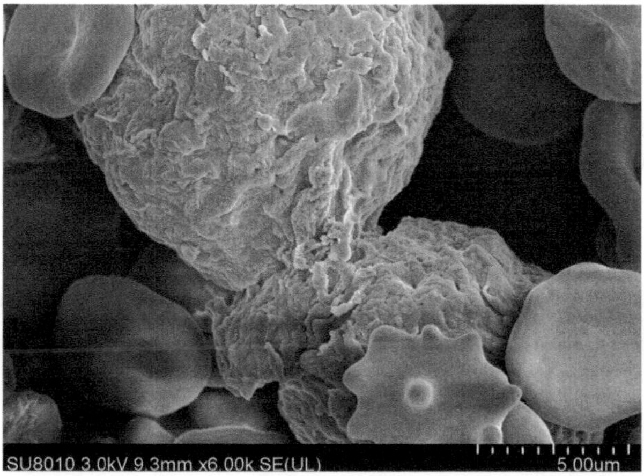

Fig. 3.9 Scanning electron microscopy of tightly connected adenocarcinoma cells

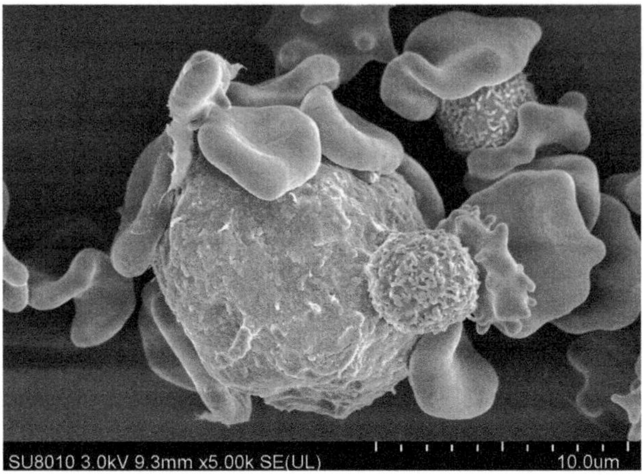

Fig. 3.10 Interaction between adenocarcinoma cells and macrophages under scanning electron microscopy
Macrophages are often captured by tumor cells and become tools for information transmission and material transportation of tumor cells. Through exocytosis and endocytosis, the genetic information or other substances that the tumor cells are about to transmit form exoteric vesicles, which cannot be recognized by macrophages after immune camouflage. Macrophages carry these exoteric vesicles to other tissues or body fluids and release them to other tumor cells, thereby completing the transmission of information or the transfer of substances

3.1.2 Adenocarcinoma of the Peritoneum

Case: Patient, female, 65 years old, chronic course of disease. The case had peritoneal cake-like thickening and intestinal adhesions as clinical manifestations, adenocarcinoma cells were found in ascites, with no primary lesions in other organs. Tumor cells in the abdominal cavity are usually shed from tumors in neighboring organs or metastasized through blood and lymphatic vessels. But this patient was special. B-ultrasound and PET-CT general examination did not find tumor tissue in other organs, only cake-like hyperplasia was found in the peritoneum, and the tracer distribution was increased. Morphological examination of body fluid cells was shown in Fig. 3.11, 3.12, 3.13, 3.14, and 3.15.

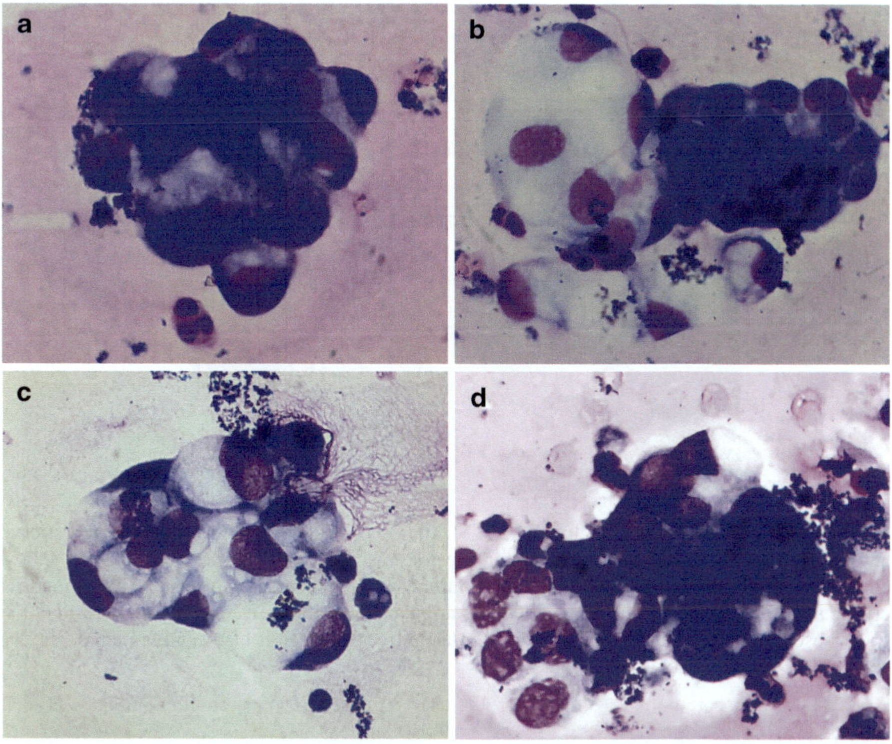

Fig. 3.11 Adenocarcinoma cells (Ritter-Giemsa staining, 1000×)
Adenocarcinoma cells often form clusters that display an irregular arrangement, hyperchromatic nuclei, and a "smudged" appearance. The secretion of mucin by some tumor cells can cause dye particles to readily adhere, which should not be misinterpreted as an artifact of poor slide preparation

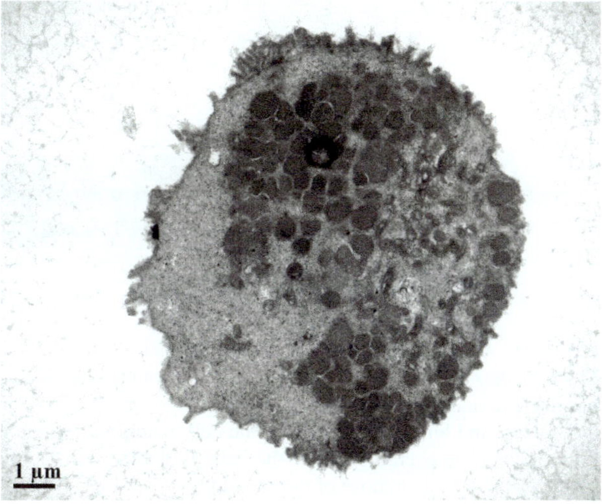

Fig. 3.12 Adenocarcinoma cells (TM, 12,000×)
Adenocarcinoma cells are rich in acinus and mitochondria, with a large number of free ribosomes, microvilli are distributed in clusters on the surface of the cell, and the nucleus is not visible in this section, and the number and distribution of organelles in the cytoplasm reflect the characteristics of tumor cells

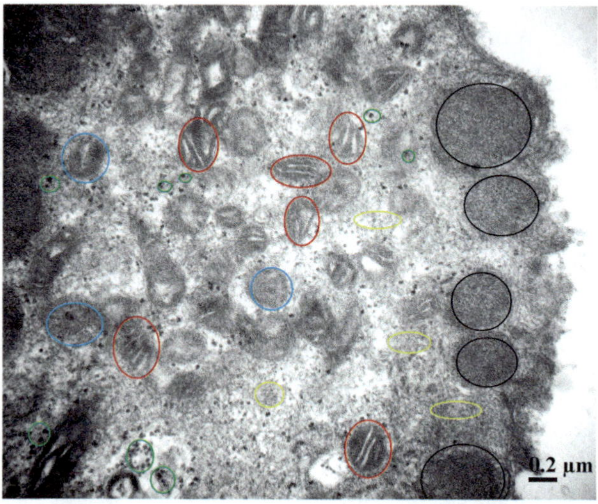

Fig. 3.13 Local magnification of Fig. 3.12 (TM, 40,000×). Mitochondria are shown as red circles, autophagic vesicles as blue circles, secretory vesicles as black circles, free ribosomes as yellow circles, and glycogen as green circles. A large number of mitochondria can be seen, mitochondrial cristae are mildly swollen, autophagic vesicles are abundant, secretory vesicles are dark in color, free ribosomes are gray and covered with cytoplasm, free glycogen is black and scattered, and the cell body is grayish-black

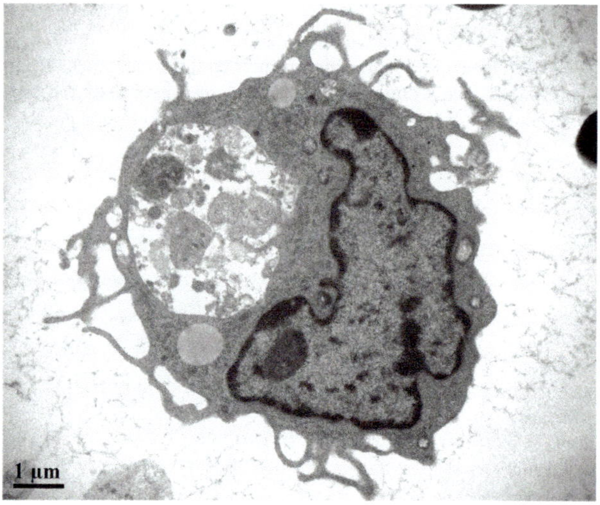

Fig. 3.14 Activated macrophages (TM, 15000×)
Macrophages are common in body fluids and should be distinguished. The nucleus is irregular and has nucleoli and many euchromatin, huge phagocytic vacuoles and residues in the cytoplasm, rich lysosomes, and long pseudopodia in the cell membrane, which is different from the microvilli of tumor cells. The overall color of the cells is lighter, and the mitochondria, free ribosomes, and glycogen are less than those of tumor cells

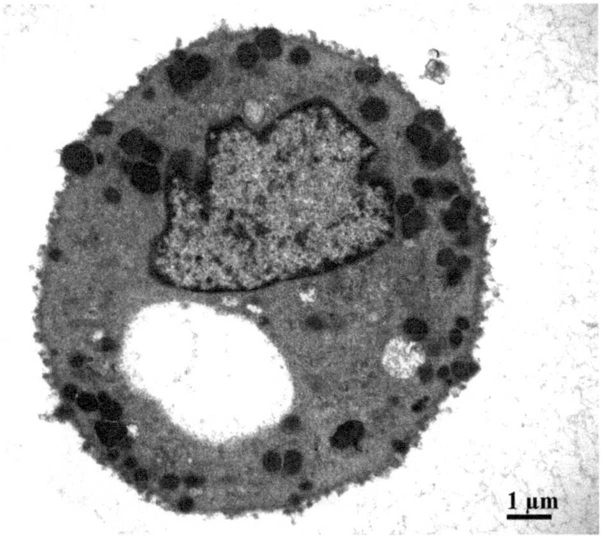

Fig. 3.15 Adenocarcinoma cell (TM, 12,000×)
The nucleus is malformed, there are large autophagic vacuoles in the cytoplasm, the secretory vesicles are numerous and dark in color, the abundant free ribosomes make the cell body deepen in color, and the microvilli on the cell membrane are short and dense

3.1.3 Colorectal Adenocarcinoma with Lung Metastasis

Case: Patient, male, 75 years old, chronic course of disease. The clinical diagnosis included varicose veins of both lower limbs with phlebitis, thrombophlebitis, post-operative colon cancer chemotherapy (stage pT3N1M1), secondary mediastinal malignancy, secondary clavicular lymph node malignancy, pulmonary nodules, and pleural effusion. Morphological examination of body fluid cells is shown in Figs. 3.16, 3.17, 3.18, 3.19, 3.20, and 3.21.

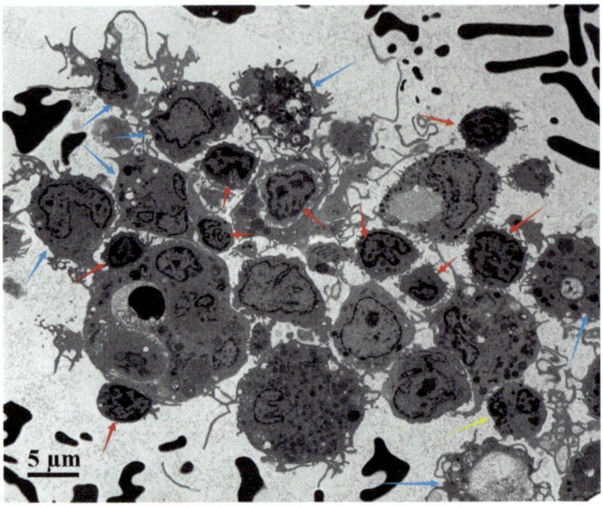

Fig. 3.16 Cell mass (TM, 3000×)
The red arrow indicates lymphocytes; the blue arrow indicates macrophages; the yellow arrow indicates neutrophils; and the rest are adenocarcinoma cells. The figure shows that adenocarcinoma cells are rich in phagocytic vacuoles, secretory vesicles, and autophagic vacuoles, as well as phagocytosis of lymphocytes. Tumor cells can be seen in large acini with short and dense villi; in the middle of the figure, several poorly differentiated tumor cells have large nucleo-cytoplasmic ratio, obvious nucleoli, and abundant euchromatin. Tumor cells often appear in clusters with other cells; note the morphological characteristics of different cells, neutrophils, lymphocytes and peripheral blood of similar size; activated macrophages increased in size, different sizes; well-differentiated tumor cells are larger in size; and poorly differentiated tumor cells are smaller in size

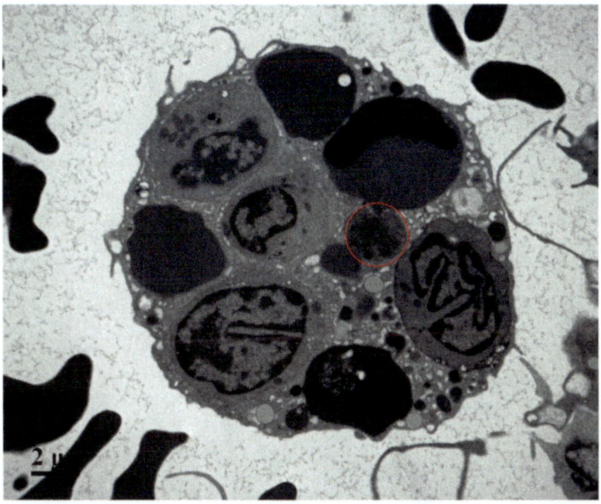

Fig. 3.17 Adenocarcinoma Cells (TM, 7000×)
Tumor cells are prone to phagocytosis, black dense round cells are erythrocytes phagocytosed by tumor cells, and the remaining nucleated cells are lymphocytes. Typical lymphocytes in body fluids have round or round-like nuclei, are easily activated, and show an increase in reactive lymphocytes, with lobulated nuclei, more heterochromatin, and fewer organelles in the cytoplasm, mainly mitochondria. Reactive lymphocytes with lobulated nuclei are easily confused with macrophages, which have more euchromatin than lymphocytes, less heterochromatin, more cytoplasmic organelles than lymphocytes, mainly lysosomes, larger cell body, and obvious and elongated pseudopods. Tumor cells synthesize and proliferate exuberantly, which can be seen in the red circle in the picture, and large aggregates of glycogen are rare in normal cells. A large number of free ribosomes make the cells gray-black, which is also one of the characteristics of tumor cells

Fig. 3.18 Tumor cells (TM, 8000×)
The red arrows are phagocytosed small tumor cells with unclear structure and a large number of acinus; the yellow arrows are lymphocytes. Because of the excessive growth of tumor cells, phagocytosis often occurs, not only phagocytosis of similar cells but also phagocytosis of other cells

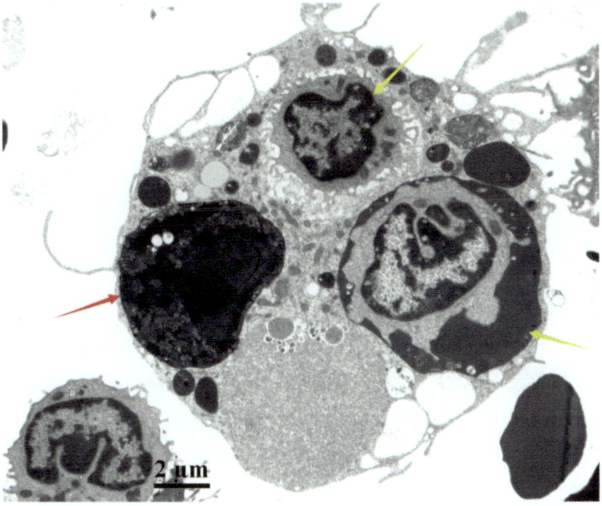

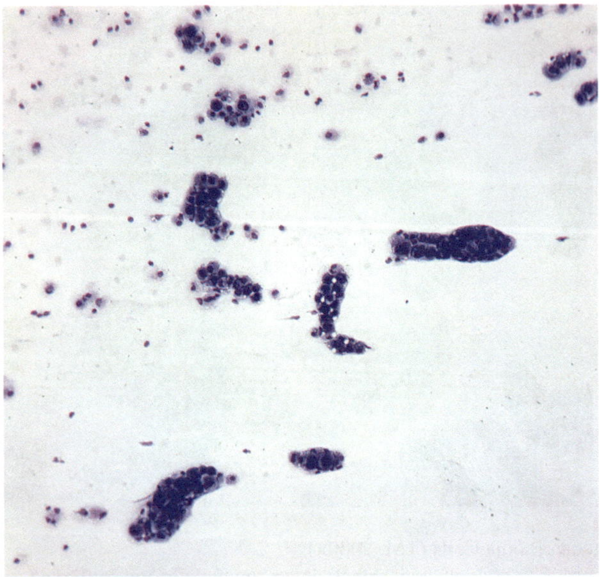

Fig. 3.19 Tumor cell mass under low magnification (Ritter-Giemsa stain, 100×)
To analyze the morphology of body fluid cells, it is necessary to first take a general view of the whole film under low magnification, and the field of view is large, and when you see cancer clusters or cells with large cell body and deeply stained nuclei, you need to turn to oil microscopy for confirmation. Large cell clusters are not easy to be missed, while less differentiated, smaller, and scattered cancer cells are easy to be missed, and the morphological characteristics of each cell should be carefully observed. The observation of cell structure and internal organelles by electron microscopy is more adequate, but the requirement of specimen retention is higher, and the cells will be easily lost in the process of filming, which is similar to the embedded filming of pathology, and the advantage of screening tumors by electron microscopy is not obvious; therefore, it is more meaningful to use electron microscopy to support the evidence in cases with suspected tumors detected by light microscopic morphology analysis

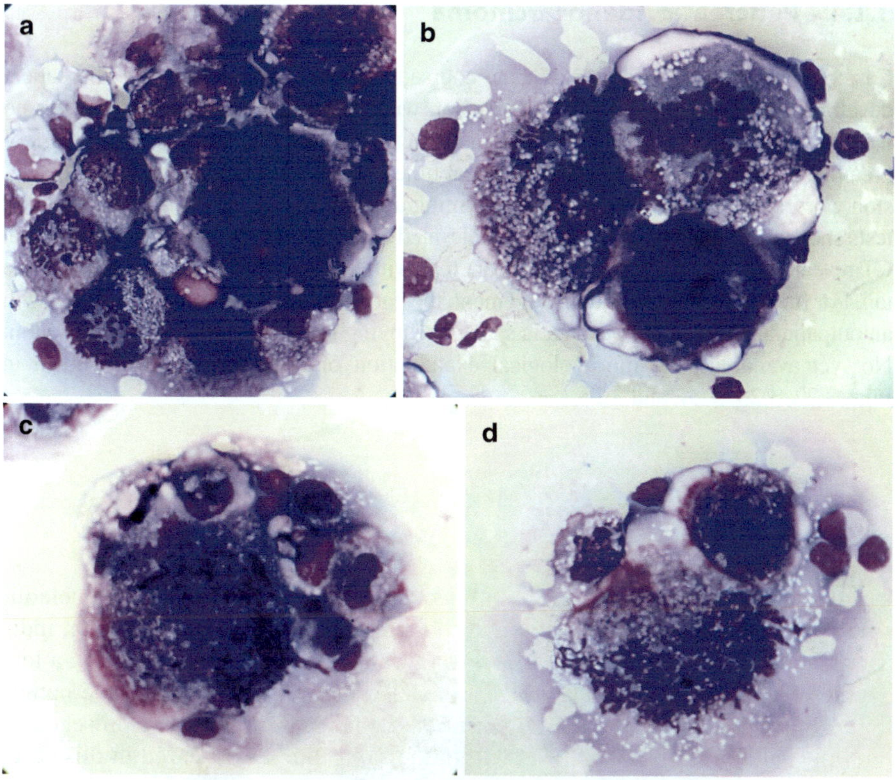

Fig. 3.20 Mitotic figure of tumor cells in malignant fluid (Ritter-Giemsa stain, 1000×)
The mitotic figure of tumor cells in malignant effusion is characterized by a high number of chromosomes, disorganized and uneven distribution, and aneuploid or hyper twofold mitotic cells, which indicates abnormal regulation of growth and development of the tumor cells and chaotic proliferation, suggesting that a tumor may be present, and the diagnosis needs to be confirmed by combining with the clinical symptoms, imaging, and pathology results. Pathologic mitotic figures are graded and evaluated together with the Ki-67 index. The more mitotic figures per high magnification field of view, the higher the degree of malignancy and the worse the prognosis

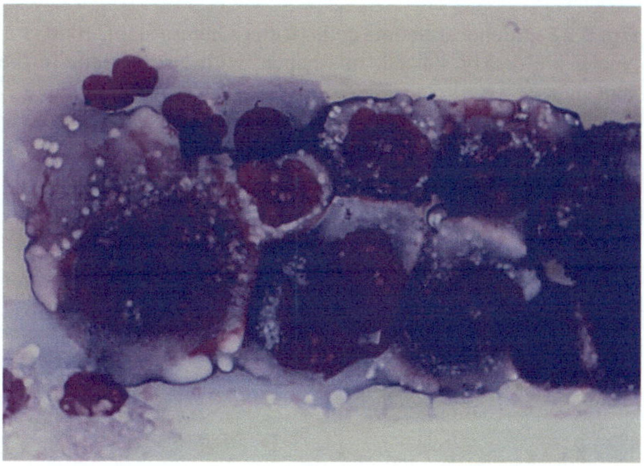

Fig. 3.21 Clustered tumor cells (Ritter-Giemsa stain, 1000×)
Tumor cells appeared in clusters, with obvious acinus in the cytoplasm, and phagocytosis was visible, making it difficult to clarify the properties of the contents, which was more clearly observed using electron microscopy

3.1.4 Pancreatic Adenocarcinoma

Case: Patient, male, 82 years old, acute course of disease. One month ago, the patient developed postprandial abdominal distension without an obvious cause. There was no nausea, vomiting, abdominal pain, dizziness, headache, chest tightness, or shortness of breath. The condition was not treated at that time. One week ago, the abdominal distension worsened, occurring even after consuming a small amount of water. Tumor marker tests showed CA19-9 at 1000 U/mL and elevated CA125. Contrast-enhanced abdominal CT revealed a mass in the body of the pancreas with associated atrophy of the pancreatic tail. For further evaluation and management, the patient was admitted to the hospital with an outpatient diagnosis of pancreatic space-occupying lesion. Pathological examination: Not yet available. The morphological examination of body fluid cells is shown in Figs. 3.22, 3.23, 3.24, and 3.25.

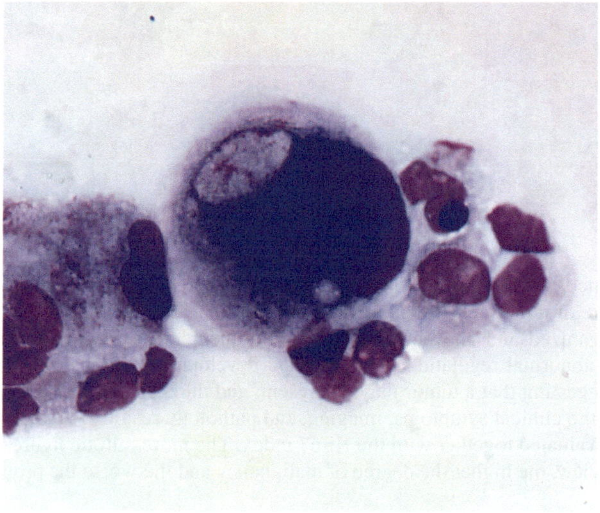

Fig. 3.22 Adenocarcinoma cells (Ritter-Giemsa stain, 1000×)
Nuclei are darkly stained, misshapen, with large acinus in the cytoplasm, and overall cloudy appearance

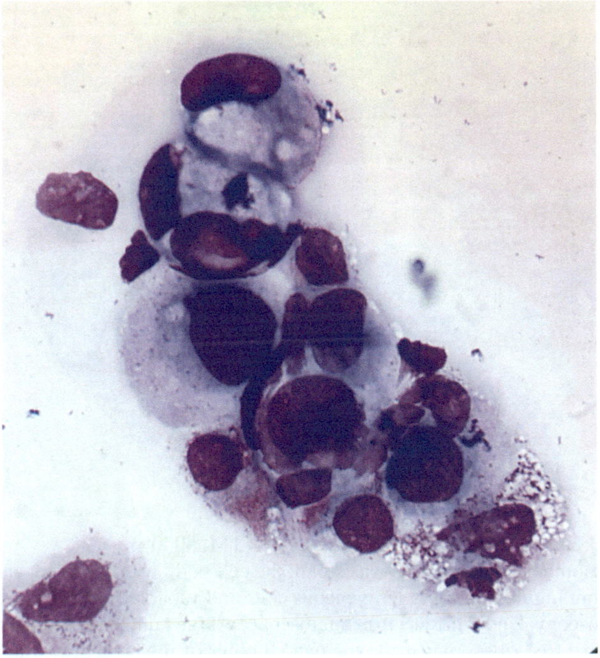

Fig. 3.23 Adenocarcinoma cells arranged in clusters (Ritter-Giemsa stain, 1000×) Nuclei malformed, cloudy cytoplasm, large secretory vesicles are seen

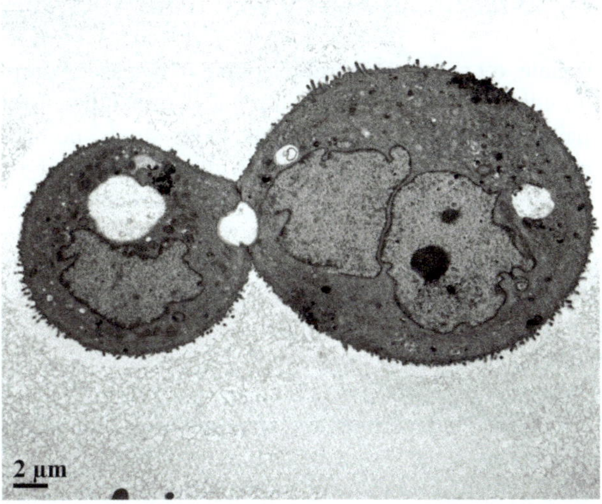

Fig. 3.24 Adenocarcinoma Cells (TM, 5000×)

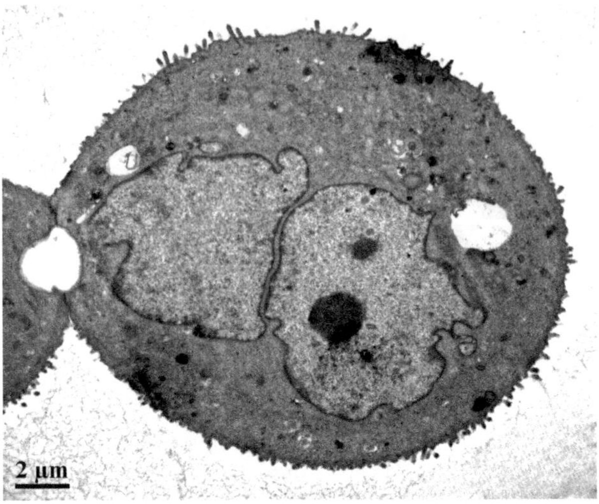

Fig. 3.25 Fig. 3.24 Magnified adenocarcinoma cells (TM, 8000×)
In this case, the number of tumor cells was small, the cell body was not large, and it was not easy to distinguish from highly nuclear heterogeneous cells and vacuolar-like change mesothelial cells under light microscopy. Under the electron microscope, a large number of secretory vesicles, heaps of glycogen, dense free ribosomes made the overall color of the cells gray-black, nuclear deformity, large nucleoli, abundant mitochondria, euchromatin nuclei, and a large number of short and dense villi on the surface of the cells were seen. The differences in morphology with mesothelial cells were obvious, reflecting the advantages of electron microscopic morphology analysis

3.1.5 Colorectal Mucinous Adenocarcinoma

Case: Patient, female, 69 years old, chronic course of disease. Admitted to the local hospital for "1 month of bloating." At that time, abdominal distension was obvious, no abdominal pain, no diarrhea, bloody stools, black stools, no nausea and vomiting, and no fear of cold and fever; abdominal ultrasound report suggested "large effusion in the abdomen-pelvic cavity." Chest CT non-contrast scan report suggested the following: ① ground glass nodules were found in the upper lobe of the right lung, lung tumor awaiting expulsion; ② calcification foci in the lower lobe of the left lung, local thickening of the pleura on both sides; ③ attached intrahepatic low-density shadow, ascites, right upper abdominal fat space to see nodular shadow. Total abdominal spiral CT enhancement report suggested the following: ① transverse colon wall thickening, considering tumor, colonoscopy was recommended; ② thickening of omentum and peritoneum with nodules, tumor metastasis to be excluded; ③ multiple low-density opacities in the liver, some cysts metastasize to be excluded, and MRI enhancement was recommended; ④ after gallbladder surgery, both kidney cysts. For further treatment, colonoscopy was performed, and the report suggested the following: MT of the transverse colon, multiple polyps in ascending colon, with the thickness of the echoless zone (CM) was 5.5 in the right lower abdomen, and the thickness of the echoless zone (CM) was 5.0 in the left lower abdomen. The high echo area of fiber strip was observed. Colonoscopy pathology report suggested the following: villous adenoma of transverse colon mucosa (transverse colon) with moderate dysplasia of glandular epithelium. The specific diagnosis and treatment was unknown and untreated. One month later, she developed abdominal pain due to external gravity pressure. She went to the local hospital again and underwent abdominal CT examination. The report suggested that thickening of the omentum and peritoneum with multiple nodules were considered metastases. The ascending colon wall thickened, liver with multiple low-density shadows, gallbladder postoperative changes, pelvic floor fascia thickened. The initial diagnosis was considered "malignant tumor." Figures 3.26, 3.27, 3.28, 3.29, 3.30, 3.31, 3.32, and 3.33 show the morphological examination of body fluid cells.

Mucous filaments play a certain role in the pathogenesis and prognosis of mucous adenocarcinoma. Studies have shown that the higher the amount of mucus in cancer cells, the higher the degree of malignancy. The more mucus outside the cancer cells, the lower the degree of malignancy, and rarely the metastasis of lymph nodes, which may be related to the cancer cells floating in the rich mucus and unable to contact the mesenchyme, thereby preventing the metastasis and infiltration of tumor cells. Most researchers believe that the production and amount of mucus in tumor tissues play an important role in influencing the prognosis. Studies have pointed out that the formation of mucus causes the degeneration and necrosis of cancer cells, thus reducing the malignancy degree. The more external mucus secreted by mucinous cancer, the better the prognosis. However, mucinous adenocarcinoma needs to be distinguished from other benign diseases such as fibroadenoma, papilloma, and duct hyperplasia. Metastatic mucinous adenocarcinoma should be examined by B-ultrasound, radiation, and gastroenteroscopy to exclude tumors of digestive tract and reproductive tract.

Fig. 3.26 Mucinous cell clusters (Ritter-Giemsa stain, 100×)
Under the low magnification microscope, there were many cell clusters adhering to the fibrous filaments, the cells were dark in color, distributed in clusters, and the structure was not clear

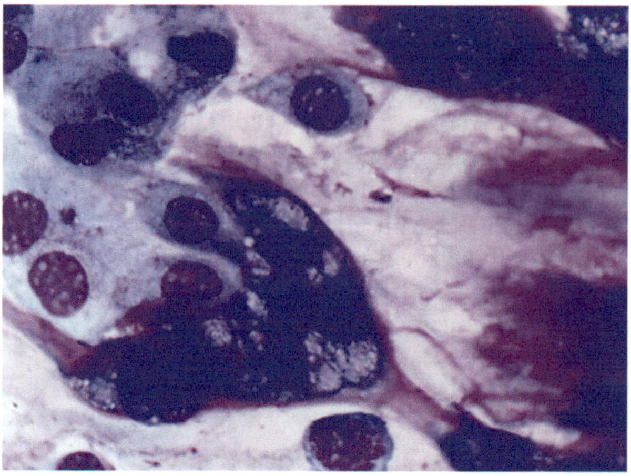

Fig. 3.27 Tumor cells wrapped in fibrous filaments (Ritter-Giemsa stain, 1000×)

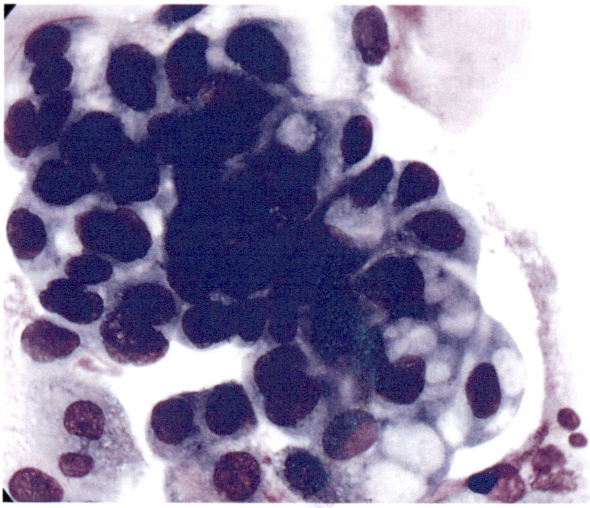

Fig. 3.28 Clustered adenocarcinoma nests (Ritter-Giemsa stain, 1000×)

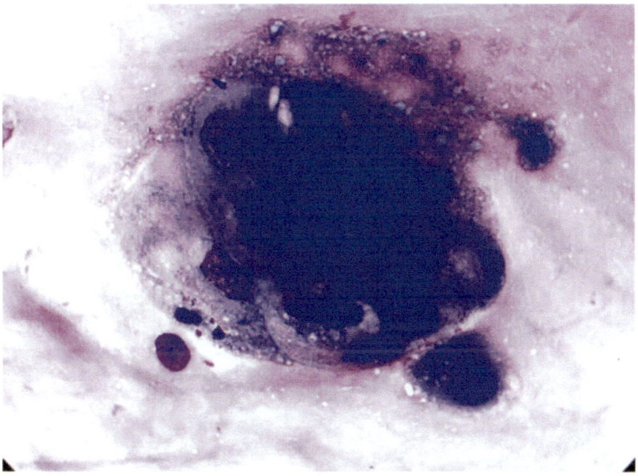

Fig. 3.29 Adenocarcinoma nest wrapped by fibrous filaments (Ritter-Giemsa stain, 1000×)
Due to the wrapping of a large number of mucus filaments, the cell clusters were not clear, and at the edge of the cell clusters, more typical and clear tumor cells could be seen. Due to the high mucus secretion, the morphology is difficult to recognize under the light microscope

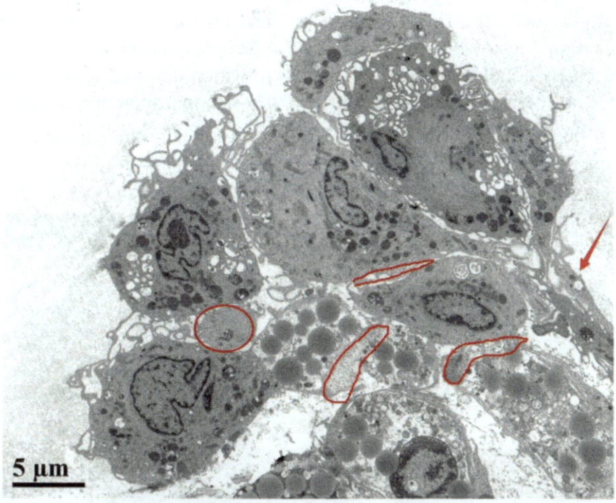

Fig. 3.30 Mucous adenocarcinoma nest (TM, 4000×). Mucous filaments are indicated by the red arrows and in the red circles

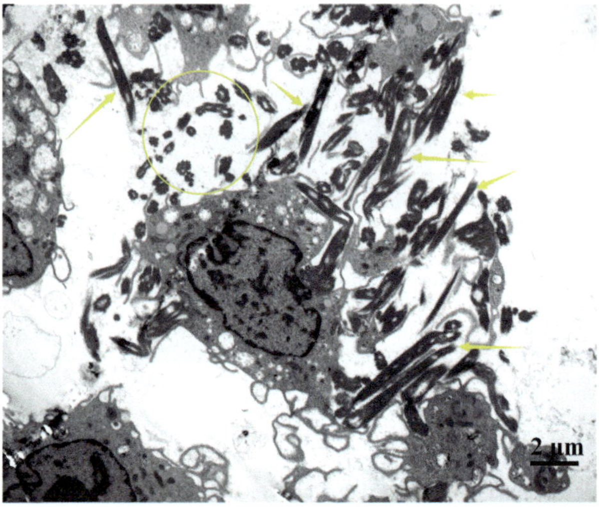

Fig. 3.31 Tumor cells wrapped by mucus (TM, 7000×)
The yellow arrows point to mucus filaments, and the yellow circle is the cross section of mucus filaments. It can be clearly seen that the mucus filaments are relatively dense and disordered under the electron microscope, and they encapsulate the tumor cells, and in Fig. 3.30, a mass of tumor cells was gathered into a pike shape under the action of mucus filaments, which was consistent with the morphology under the light microscope. The characteristics of mucinous adenocarcinoma: the mucous filaments are abundant, the cells appear in clusters and are arranged tightly, and the dye residue is easily adhered to the cell clusters in Ritter staining. It is not that the chromatographic staining is not cleanly rinsed out, but it is the strong adherence of the specimen, which makes the dye residue adhered to the mucous filaments

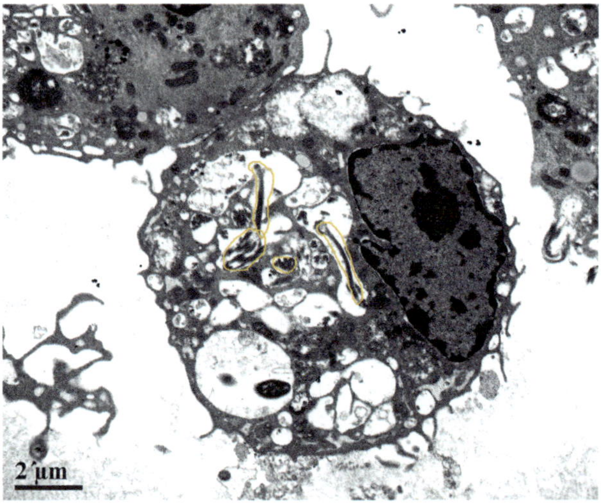

Fig. 3.32 Tumor cells (TM, 10,000×) with a large number of phagocytosed mucus filaments in the cytoplasm (yellow circle)

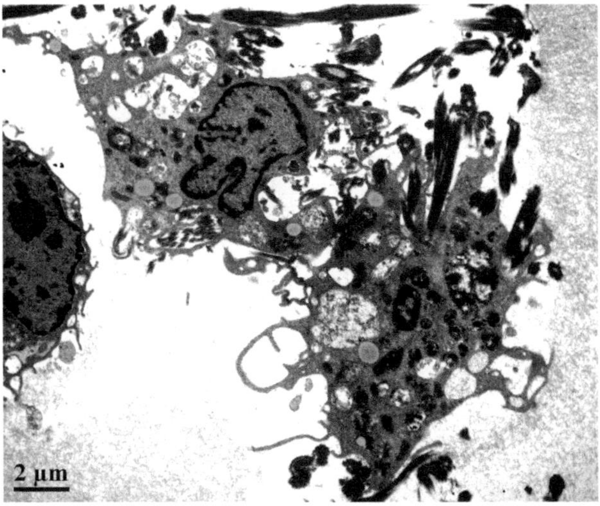

Fig. 3.33 Adenocarcinoma cells (TM, 8000 ×)
Two tumor cells are held together by mucus filaments, and phagocytosed mucus filament inclusion bodies are visible in the cytoplasm; the long one is a longitudinal section and the round one is a transverse section

3.1.6 Gastric Signet-Ring Cell Carcinoma, Peritoneal Adenocarcinoma

Case: Male, 82 years old, with acute and chronic disease, was admitted to the hospital 4 months ago with "abdominal distension for 3 months and difficulty in eating for half a month." Gastroscopic pathology report suggested "gastric body" low adhesive carcinoma (signet-ring cell carcinoma). PET-CT report suggested diffuse irregular thickening of gastric wall in the gastric body; diffuse increase of FDG metabolism, which was consistent with the imaging manifestation of gastric cancer (leathery stomach); and perigastric and retroperitoneal abdominal aorta right lymph node metastasis. The patient had peritoneal implantation metastasis and malignant fluid in the abdominal and pelvic cavities. Patient with gastric cancer and multiple lymph node metastasis experienced no discomfort such as nausea and vomiting after chemotherapy, and abdominal CT examination report suggests peritoneal implantation metastasis and large amount of ascites. The morphological examination of body fluid cells is shown in Figs. 3.34, 3.35, 3.36, 3.37, and 3.38.

Signet-ring cell carcinoma (SRCC) is a highly malignant mucus-secreting adenocarcinoma originating from the mucosal epithelium, in which the cancer cells produce mucin and the cytoplasm is filled with mucus, which squeezes the nucleus on one side of the cytoplasm. It is named after its "signet-ring" appearance under light microscope. It is most common in primary foci in the gastrointestinal tract but may also occur in the breast, gallbladder, bladder, and pancreas. Unlike the previous case where the mucus was extracellular, in this case, the mucus was intracellular, with high degree of malignancy, had infiltrative growth, was easy to diffuse and metastasize, had poor chemotherapeutic effect, had good early surgical effect, and has poor prognosis in late stage.

Fig. 3.34 Clustered adenocarcinoma cells (Ritter-Giemsa Stain, 1000×)

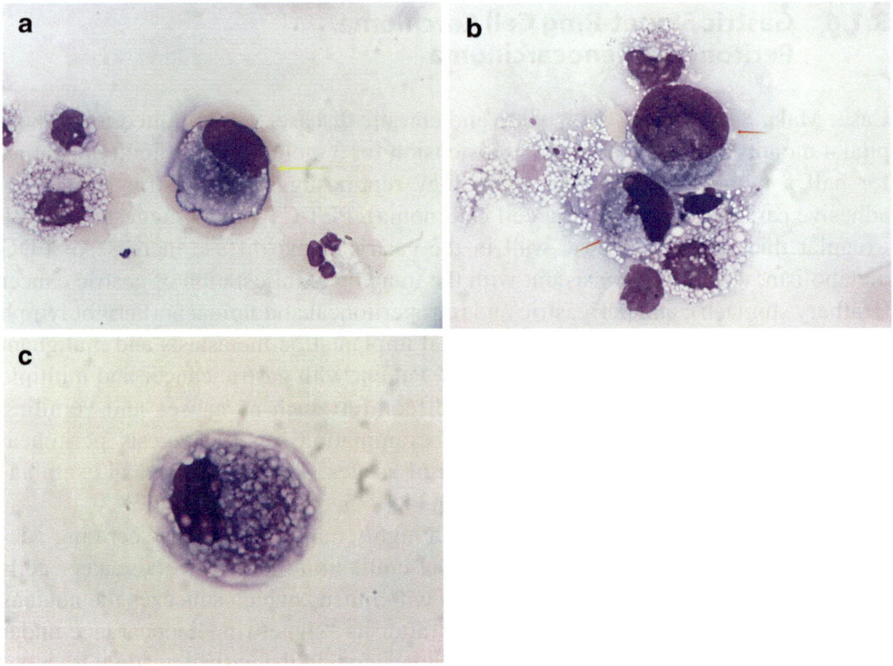

Fig. 3.35 signet-ring cell (Ritter-Giemsa stain, 1000×)

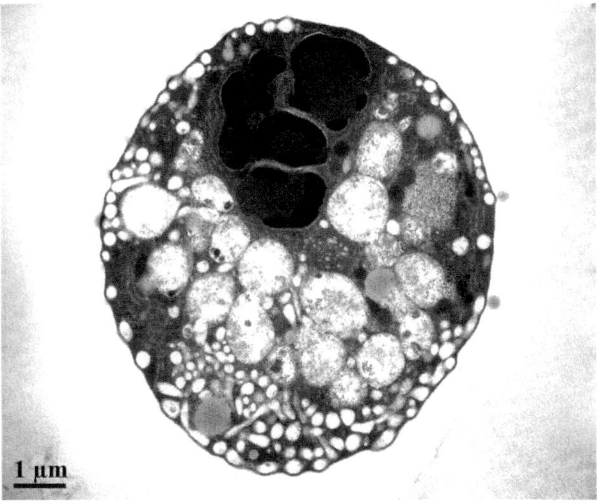

Fig. 3.36 Adenocarcinoma cells (TM, 15,000×)
The cytoplasm was full of mucus-containing acinus, and a large number of free ribosomes, endoplasmic reticulum, and mitochondria could be seen in the empty space, suggesting that the cell synthesized and secreted exuberantly. The nucleus of the cell has partial karyopyknosis, which is a sign of impending apoptosis

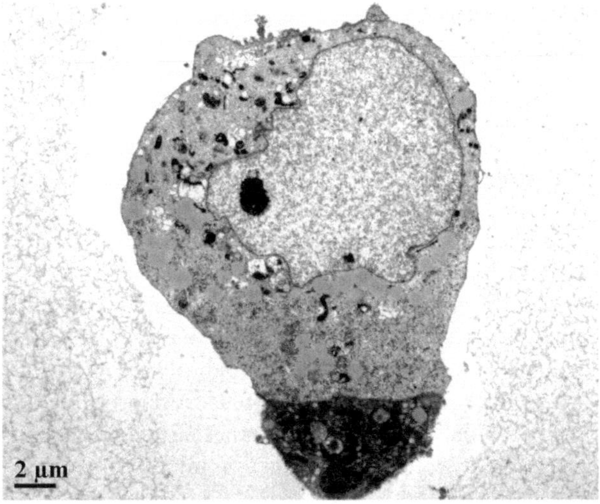

Fig. 3.37 Tumor cells (TM, 6000×)

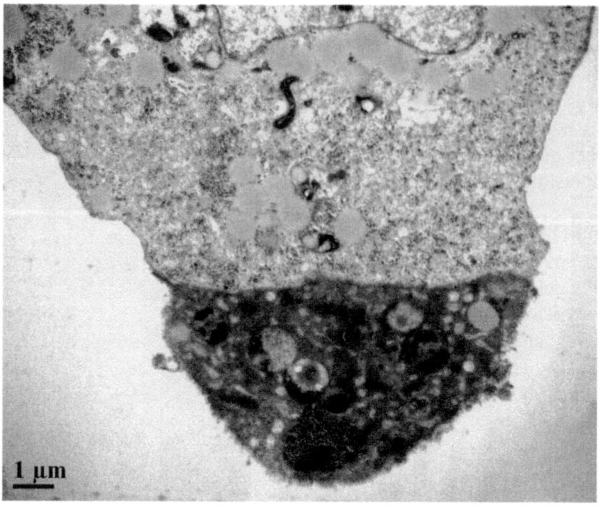

Fig. 3.38 Fig. 3.37 Local enlargement (TM, 12,000×)
As shown in Fig. 3.37, the cell body of the two tumor cells is closely connected, with different cell body coloring and organelles, and the cell membrane can be seen vaguely at the connection of the cells. The nucleus of this large tumor cell is euchromatin-rich, and the nucleolus is obvious, but the chromatin is loosely distributed with more white, suggesting that the cell is going to be apoptotic; the cytoplasm of the other cell has many acinar vesicles with small volume, suggesting that the cell is relatively naive and not yet mature. Signet-ring cell carcinoma is a poorly differentiated carcinoma

In this case, the stomach is signet-ring cell carcinoma and the peritoneum is adenocarcinoma. Under the light microscope, the cytoplasm was rich in acinar vesicles of different sizes, the nucleus was obviously displaced, and vesicular shadows could be seen on the nucleus, which was related to the fact that the cells were three-dimensional, the cytoplasm was densely populated with acinar vesicles, and some cytoplasmic acinar vesicles would overlap on the nucleus when it was pushed to become a flat surface. The internal structure and distribution of the cells can be visualized by transmission electron microscopy.

3.1.7 Gastric Highly Differentiated Adenocarcinoma

Case: Patient, male, 82 years old, chronic course of disease. A month ago, abdominal pain and distension occurred without obvious inducement, which was paroxysmal, and the degree was not dramatic and was tolerable, accompanied by unsolved stool for more than 4 days and occasional exhaust. There were no nausea, vomiting, dizziness, headache, chest tightness, chest pain, chills, fever, cough, sputum, etc.

After 1 week, he went to a local hospital. Performed the whole abdomen CT scan, and the report suggested that abdominal gas and fluid accumulation and gastrointestinal perforation were possible. The tumor indexes were checked, and the results showed that carcinoembryonic antigen was 22.27 ng/ml and carbohydrate antigen 19–9(CA19–9) was 67.60 U/ml. He was discharged form hospital after treating with cefodizime for anti-infection, fasting, stomach protection, spasmolytic relief, and slight improvement. After discharge, the patient still had abdominal distension and sought medical attention. CT examination indicated that the gastric wall was thickened locally in the stomach body. Considering the neoplastic lesions, the local boundary between the body and the tail of the pancreas was not clear, and the degree of parenchymal enhancement of the pancreas was reduced. Endoscopic and enhanced MRI examination were recommended. The gastroscopy report suggested gastric cancer (repeat biopsy if necessary) or antral polyps. The pathological results indicated that "body of stomach" was highly differentiated adenocarcinoma. For further treatment, he was admitted to hospital for a preliminary diagnosis of "gastric malignant tumor." Morphological examination of humoral cells is shown in Figs. 3.39, 3.40, 3.41, 3.42, and 3.43.

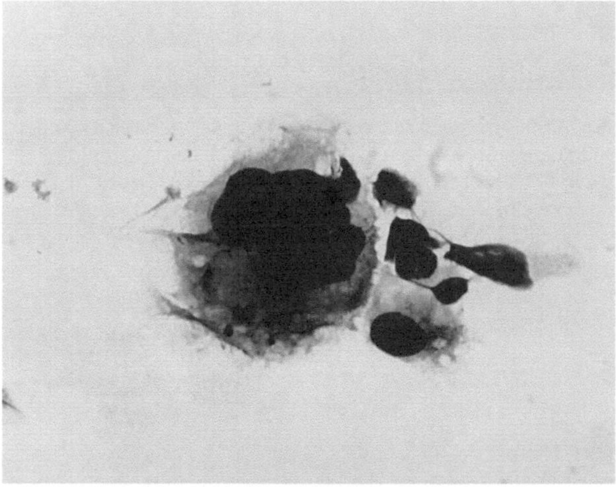

Fig. 3.39 Nuclear malformation adenocarcinoma (Ritter-Giemsa stain, 1000×)

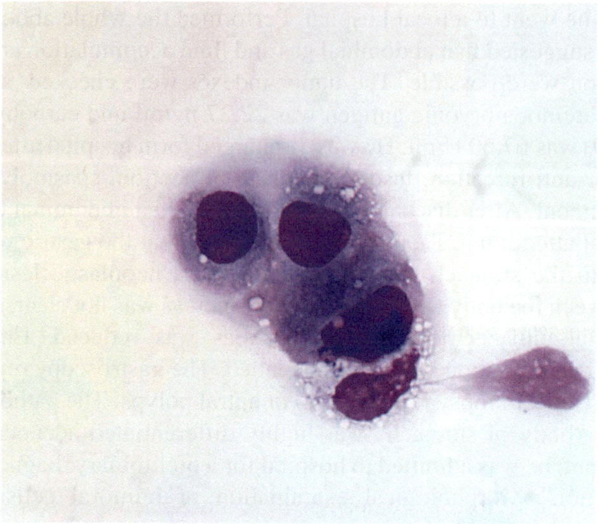

Fig. 3.40 Adenocarcinoma in small aggregates (Ritter-Giemsa stain, 1000×)

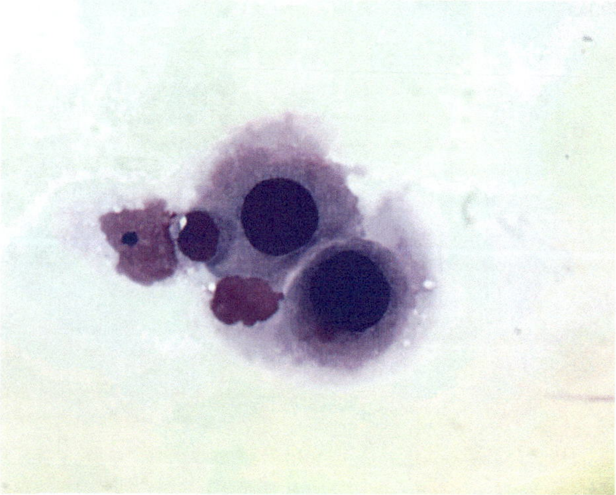

Fig. 3.41 Adenocarcinoma cells (Ritter-Giemsa stain, 1000×), resembling squamous cell carcinoma
Under the light microscope, the morphology of this cell was easily confused with squamous cell carcinoma, with no obvious acinus, and the cell body could be seen with purplish-red stained villi, and the border was not clear. Pathology confirmed the diagnosis of highly differentiated adenocarcinoma, and the morphology did not match with the pathologic diagnosis. Due to the variability of morphology, the diagnosis should be based on the pathologic conclusion

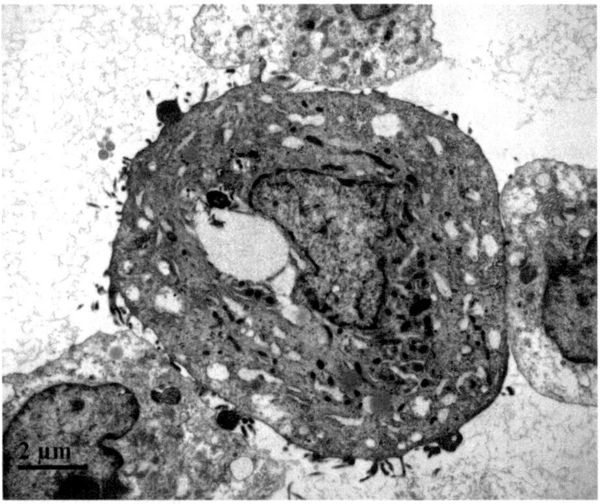

Fig. 3.42 Electron microscopy of adenocarcinoma cells (TM, 10 000×)
There are obvious microvilli around the cells, obvious secretory vesicles in the cytoplasm, which are small and numerous, mitochondria are obviously increased, free ribosomes are distributed in clusters or scattered, and euchromatin in the nucleus is obviously increased, which suggests that the cells are actively proliferating and differentiating

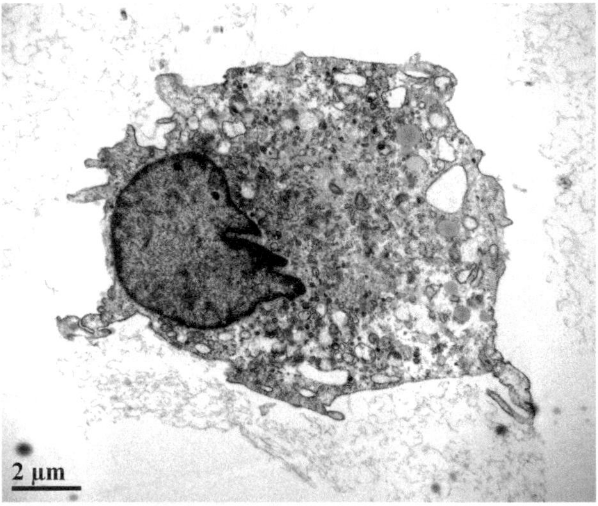

Fig. 3.43 Electron microscopy of apparently apoptotic adenocarcinoma cells (TM, 10000×)
Apoptosis is clearly characterized by fragmentation of the cell membrane, swelling and deformation of organelles in the cytoplasm, partial lysis, homogeneous changes in the nucleus, and fading of the granularity of the chromatin. Although the cells are in apoptosis, the cytoplasm is still densely covered with small secretory vesicles and free ribosomes, glycogen and mitochondria are increased, and the endoplasmic reticulum is swollen and fragmented. Body fluid cells in the body fluid shedding survival time varies, some have been shedding immersed for a long time, despite the timely delivery, as soon as possible detection, apoptotic cells can still be seen. It is difficult to distinguish whether some special cases are adenocarcinoma or squamous cell carcinoma from light microscopic morphologic analysis. Pathological immunohistochemistry and electron microscopic morphologic analysis are good means of confirming the diagnosis, and the application of these two means has also been recognized internationally

3.1.8 Postoperative Abdominal Metastasis from Colon Cancer

Case: Patient, male, 83 years old, with chronic disease, was admitted to the hospital 3 months ago with "abdominal pain and bloating for 1 week." Complete examination and exclusion of contraindications were performed laparoscopic radical resection of the right hemi-colon, intestinal adhesion release, and biopsy of the mesenteric nodes, and the patient recovered well after the operation. Pathological examination of the specimen revealed: A right hemicolectomy specimen (colorectal carcinoma) and mesenteric nodules: ① ileocecal bulge type of poorly differentiated adenocarcinoma, part of it was mucinous adenocarcinoma and signet-ring cell carcinoma, the size of the tumor was 4 cm × 3.5 cm × 3.5 cm, and the tumor was infiltrated to the subplasma membrane, with vascular involvement (+) and nerve involvement (+);②the upper and lower margins were negative; ③there was no cancer involvement in the appendix; ④cancer metastasis was seen in regional lymph nodes (9/18), including 0/7 periampullary lymph nodes of the small intestine, 9/11 periampullary lymph nodes of the colon, and multiple cancerous nodules; ⑤ a small amount of adenocarcinoma infiltration was seen in fibro-fatty tissues of the mesenteric nodules;⑥TNM staging (AJCC, 8th edition): pT3N2bMx. Immunohistochemical staining results showed HER-2 (1+), CK7 (−), CK (Pan) (+), Villin (+), SATB2 (+), SALL4 (−), Ki-67 (70%), CD34 (blood vessel +), PMS2 (+), MSH2 (+), MSH6 (+), MLH1 (+), D2–40 (lymphatic +), CDX2 (+), E-cadherin (+), EGFR (partial +), and P53 (+, 90%). The patient recovered well after surgery, the wound healed well, and he was discharged. Two weeks ago, the patient presented symptoms of abdominal pain and distension without obvious inducement, decreased appetite, difficulty in anal exhaust and defecation, obvious abdominal distension, and whole abdomen distension, so he came to the hospital for treatment and was admitted to the clinic as "incomplete intestinal obstruction." Morphological examination of humoral cells is shown in Figs. 3.44, 3.45, 3.46, 3.47, 3.48, 3.49, 3.50, 3.51, and 3.52.

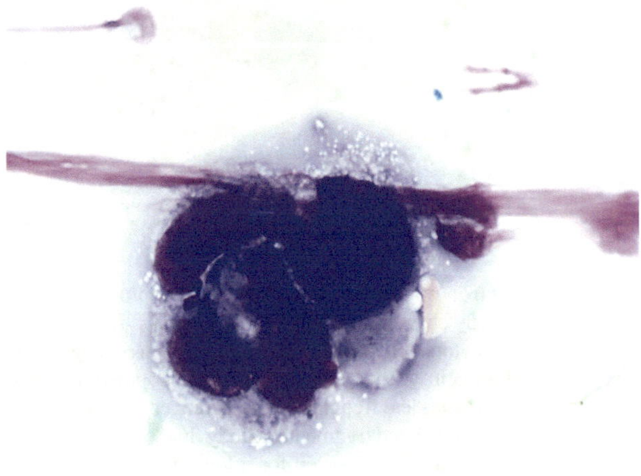

Fig. 3.44 Clustered poorly differentiated colon adenocarcinoma cells (Ritter-Giemsa stain, 1000×)

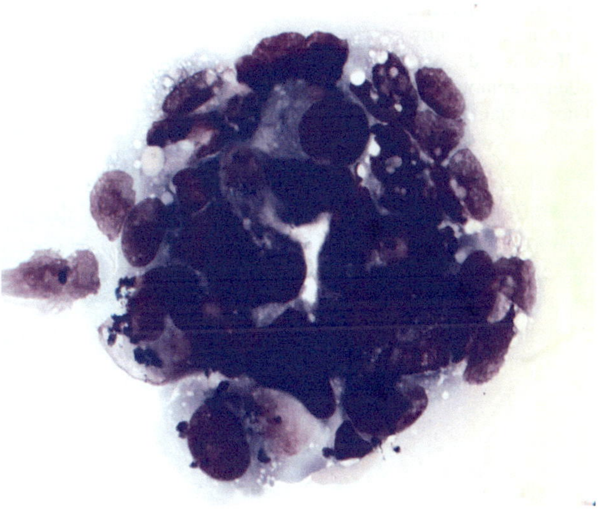

Fig. 3.45 Colonic adenocarcinoma cells with poorly defined cytoplasmic boundaries (Ritter-Giemsa stain, 1000×)

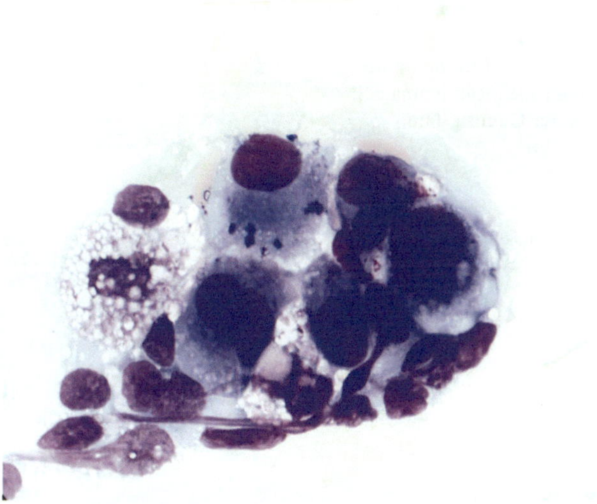

Fig. 3.46 Hypo-differentiated adenocarcinoma and signet-ring cell carcinoma cells (Ritter-Giemsa stain, 1000×)

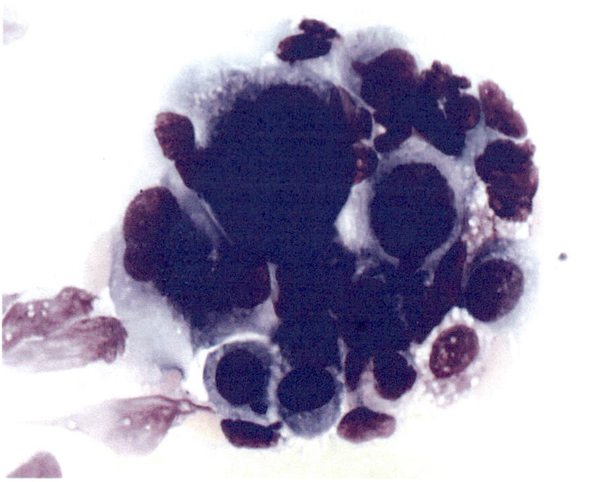

Fig. 3.47 Cancer mass with predominantly poorly differentiated adenocarcinoma (Ritter-Giemsa stain, 1000×)

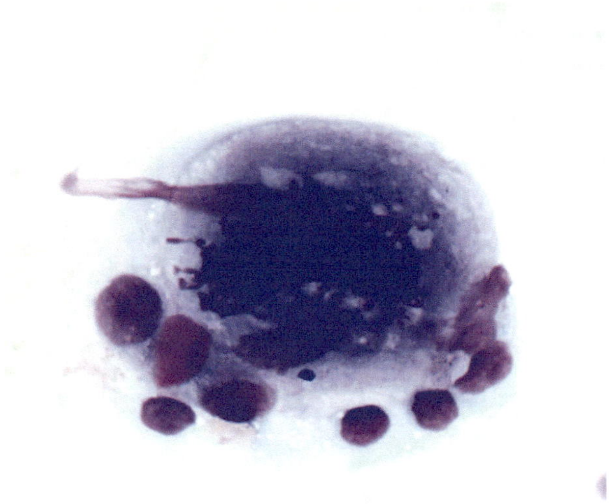

Fig. 3.48 Mitotic figure of an adenocarcinoma cell (Ritter-Giemsa stain, 1000×)

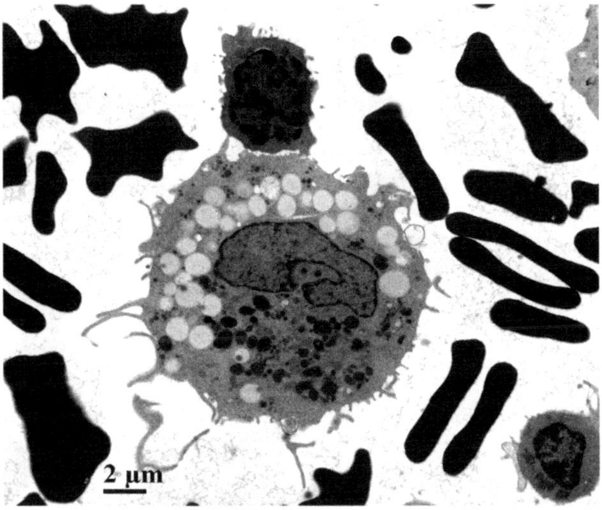

Fig. 3.49 Adenocarcinoma cells under electron microscope (TM, ×6000)
Adenocarcinoma cells are poorly differentiated, the cell body is smaller than other adenocarcinoma cells, the acinus is small and not densely packed throughout the cytoplasm, the nucleus is predominantly euchromatin, the cellular organelles are more numerous, the glycogen is scattered and less clustered, the free ribosomes are more numerous than in normal cells but fewer numerous than in mature adenocarcinoma cells, and there are pseudopods outside the cytoplasm of the cytosol of varying lengths

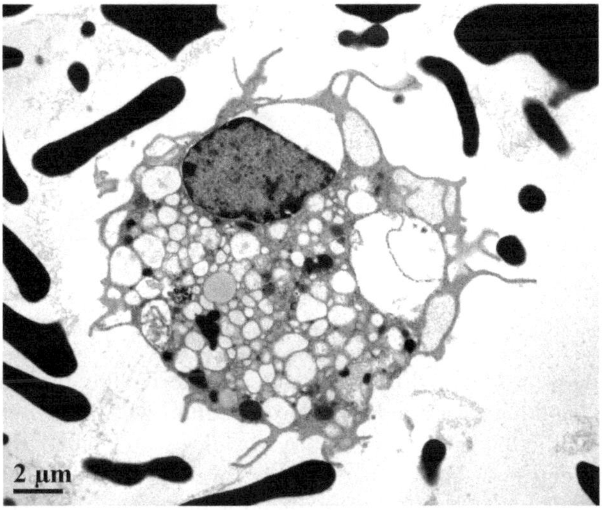

Fig. 3.50 Signet-ring cell carcinoma (TM, 7000×)
Tumor cells have medium cell body, extremely abundant acinus, varying in size, many inclusion bodies, many secretory vesicles, and the nucleus is squeezed to the edge of the cell. A large number of organelles were scattered in the cytoplasmic interstitial space, with more elongated pseudopods. The nucleus was predominantly euchromatin, and nucleoli were visible. Due to sectioning, the nuclei of different sections may be incomplete and have different morphologies, but they can be distinguished by their chromatin

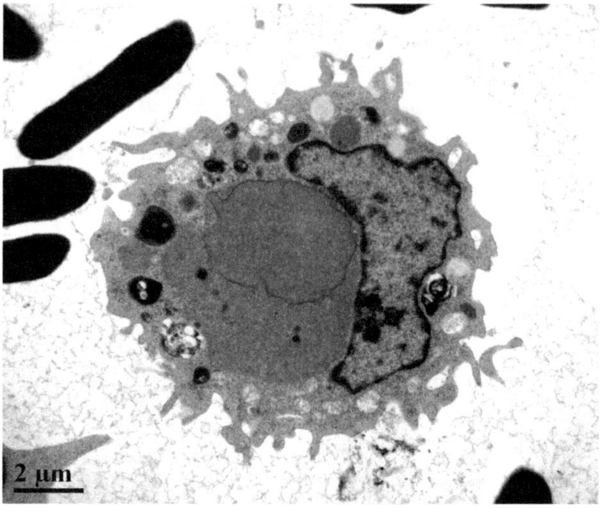

Fig. 3.51 Signet-ring cell carcinoma (TM, 10,000×) Tumor cells have large inclusion bodies, and the nuclei are pushed to one side to form signet ring-like cells, which may be composed of many small acini or a large acinus that are pushed together

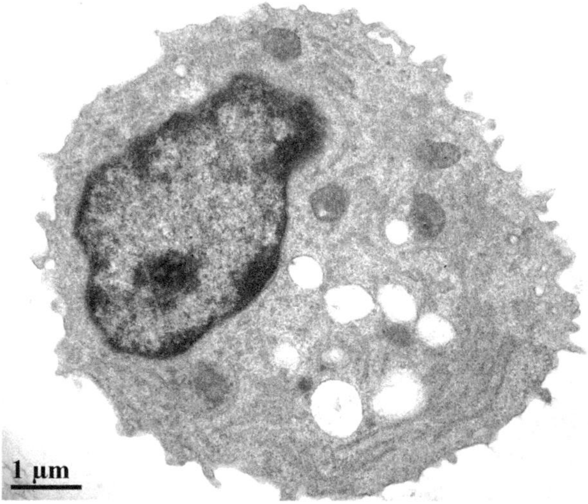

Fig. 3.52 Poorly differentiated adenocarcinoma cells (TM, 20 000×)

The tumor cells were poorly differentiated, characterized by prominent nucleoli, small cell size, and abundant endoplasmic reticulum within the cytoplasm, indicating active synthetic function. Secretary vesicles were sparse, reflecting a low degree of differentiation and limited vesicle formation, which led to the pathological classification of poorly differentiated adenocarcinoma. The presence of multiple categories of tumor cells within the same site, as observed in this case, is not uncommon. In most such tumors, one cell type predominates, while other, minor populations of different types may originate from pluripotent progenitor cells. Influenced by various factors, these progenitors can differentiate into distinct lineages, akin to how pluripotent stem cells in hematopoietic tissues give rise to different cell lines. When present in very low numbers, these non-dominant cell populations may be overlooked; however, if sufficiently numerous, they manifest as a coexistence of several tumor cell types. In general, more complex cellular heterogeneity, especially when accompanied by distant metastasis, is associated with a poorer prognosis.

3.1.9 Abdominal Metastasis of Ovarian Cancer

Case: Patient, female, 82 years old, with a chronic course of disease. There was no obvious cause of abdominal distension for 1 month, accompanied by increased abdominal circumference, and no obvious abdominal pain, diarrhea, and fever. Physical examination: sane and mentally were acceptable, no anemia, no yellow staining of the sclera. Abdomen were bloated and soft with no tenderness, rebound pain. Auxiliary examination: white blood cells count 9.09×10^9 /L, hemoglobin 110 g/L ↓, platelet count 271×10^9 /L, hypersensitive C-reactive protein 125.4 mg/L ↑, creatinine 104.4 μmol/L ↑. CT examination report showed that the abdominal fat space was blurred, with multiple flocs exudated and accumulated fluid in the abdomen and pelvis and multiple enlarged lymph nodes in the mesenteric region and retroperitoneum. Soft tissue mass appeared in uterus and bilateral adnexal area, which considered malignant tumor. The possibility of metastasis was high because emerged retroperitoneal thickening with nodules and abdominal pelvic effusion, and there were multiple enlarged lymph nodes in retroperitoneal and diaphragmatic corners of the heart. PET examination and pathology were recommended because the patient's tumor indexes were elevated and CT showed uterus and bilateral appendages' soft tissue masses, which were considered gynecologic malignancies. With abdominal metastasis, currently abdominal fluid is high, and drainage is in progress. Gynecological consultation suggested transfer surgery, chemotherapy, and other anti-tumor therapy. After communicating with the patient's family, the family considered the patient's poor tolerance and refused to transfer to another department for further treatment. At present, treatment mainly by ascite drainage reduces symptoms along with supportive treatment. Morphological examination of body fluid cells is shown in Figs. 3.53, 3.54, 3.55, 3.56, and 3.57.

A typical case of adenocarcinoma with a large number of tumor cells is usually seen in patients with end-stage tumors and suggests a poor prognosis. This patient was discharged from the hospital for maintenance therapy without any further treatment after diagnosis due to her advanced age. As the life expectancy of human beings increases, the proportion of patients with tumors is bound to increase year by year. Early diagnosis and treatment by body fluid cytomorphology can win valuable treatment time and better therapeutic effect for patients, avoiding the difficulty of treating patients who are in the terminal stage once diagnosed, which can greatly improve the quality of patients' survival and achieve a longer period of tumor-bearing survival and also save the medical expenses.

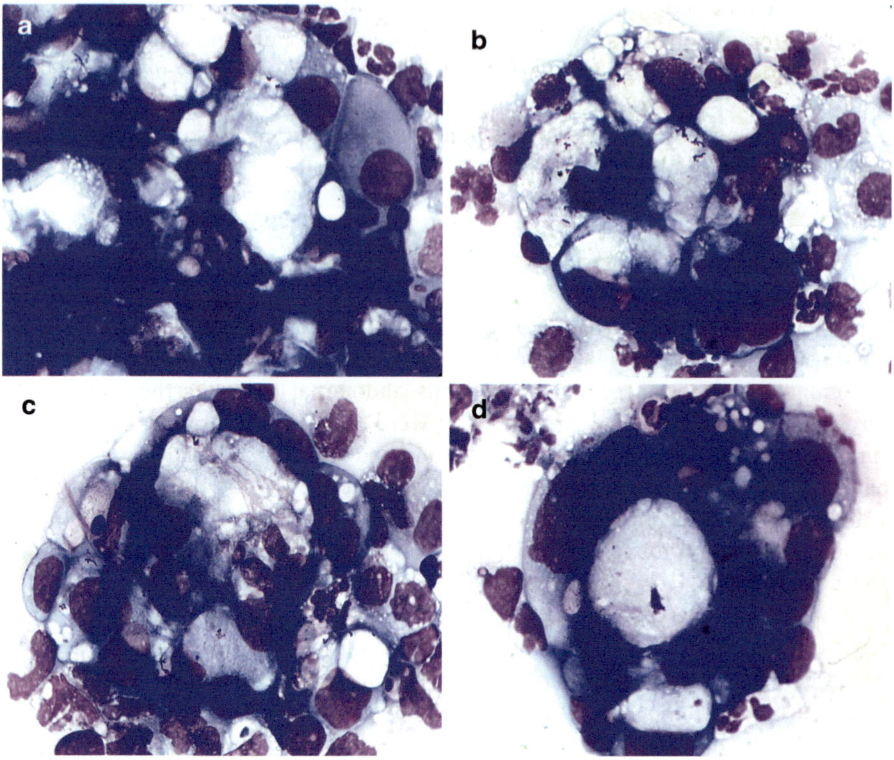

Fig. 3.53 Ovarian cancer cell cluster (Ritter-Giemsa Stain, 1000×)
This patient has extensive abdominal metastasis and lymph node metastasis of ovarian cancer, and the number of tumor cells in the ascites is high, and they are piled up into clusters. The tumor cells are densely distributed and disorganized, and "windows" can be seen at the edges of the clusters or at the areas where the cells are dispersed, and the cellular structure and abnormal morphology of the cells can be clearly seen in these areas

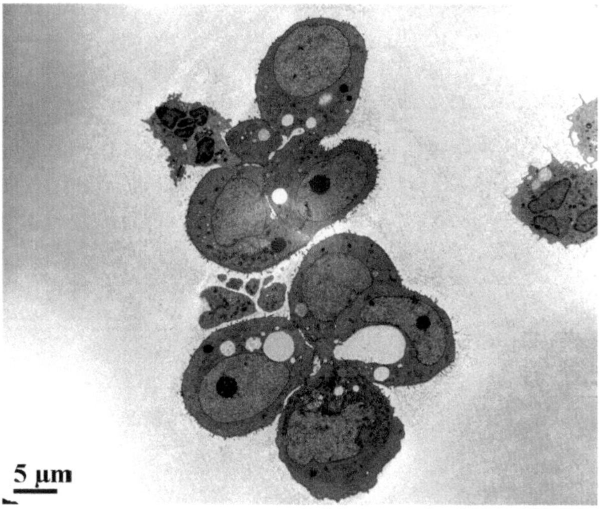

Fig. 3.54 Clustered tumor cells (TM, 2500×)
Under the electron microscope, the tumor cells were consistent with the overall morphology under the light microscope, appearing in clusters, with large cell volume, obvious nucleoli, sparse chromatin, obvious nuclear aberrations, huge acinus in the cytoplasm, abundant organelles, high number of free ribosomes, deep cell coloration, and fine microvilli distributed on the outer side of the cytomembrane

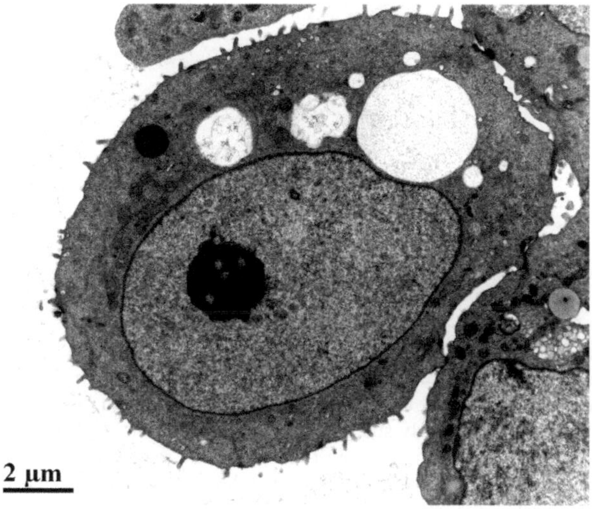

Fig. 3.55 Local enlargement (TM, 10,000×)
The tumor cells exhibited a high nuclear-to-cytoplasmic ratio, with abundant mitochondria and free ribosomes in the cytoplasm, prominent "bull's-eye" nucleoli, finely dispersed chromatin, and short microvilli

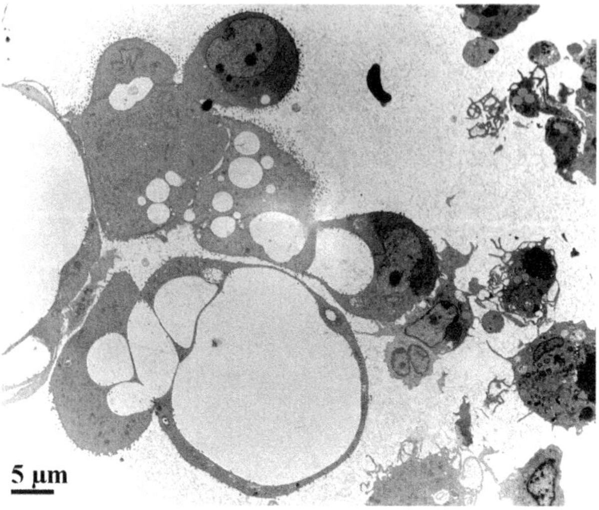

Fig. 3.56 Giant, clustered acinus (TM, 2500 ×)

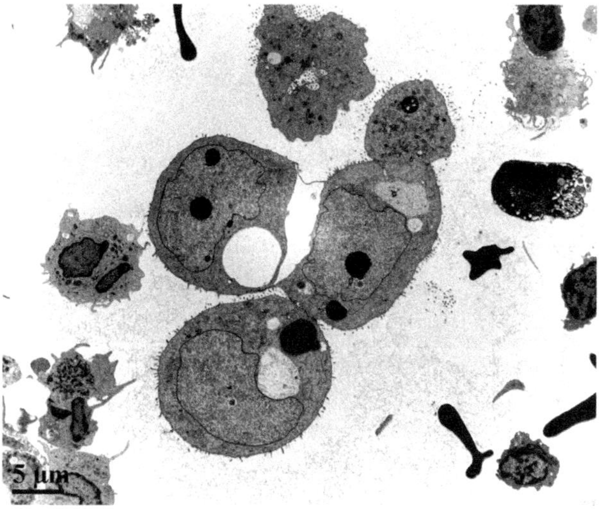

Fig. 3.57 Tumor cells with huge acinus (TM, 3000×)

Tumor cells have huge acinus, dense microvilli, large nucleoli, and large inclusion bodies. Under the electron microscope, it is easy to see clusters of tumor cells with different morphologies, and huge acinus can be seen, irregularly stacked together, and the nuclei of the cells are squeezed to one side or turned to the side so that they can't be seen. Local magnification: 1–2 large nucleoli are visible, and euchromatin is dominant; cytoplasm is less; large acinus is visible, organelles are densely packed; free ribosomes are abundant, which makes the cells deeply stained; and the aggregated glycogen is not much, which is not easy to be seen after staining; well-developed microvilli are visible outside the cell membrane

3.1.10 Moderately Differentiated Carcinoma of the Gallbladder

Case: Elderly patient, male, 71 years old, acute course. He was admitted to the hospital because of epigastric discomfort after eating. The magnetic resonance report suggested that the wall thickening at the confluence of extrahepatic bile ducts was accompanied by dilatation of the intrahepatic bile ducts, which was more likely to be inflammatory than neoplastic, but cholangiocarcinoma could not be excluded, and it was suggested that the patient be rechecked after treatment. Routine histopathological examination and diagnosis suggests "choledochal mass," moderately differentiated adenocarcinoma, 3 cm × 2 cm, and tumor infiltrating the whole layer of the wall of the choledochal duct, with nerve involvement (+) and vascular involvement (+). The morphological examination of body fluid cells is shown in Figs. 3.58, 3.59, 3.60, 3.61, and 3.62.

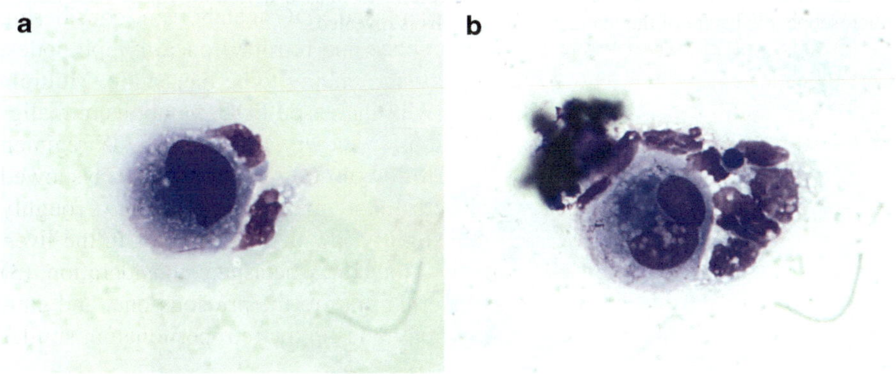

Fig. 3.58 Single, scattered tumor cells with small size, large nucleo-cytoplasmic ratio, and small, dense vesicles (Ritter-Giemsa stain, 1000×)
Scattered, single cells, distinctly heterogeneous but difficult to distinguish from mesothelial cells with moderate to high nuclear heterogeneity, which are usually mucinous and to which the dye residue readily adheres; the diagnosis is confirmed by finding a typical cell mass

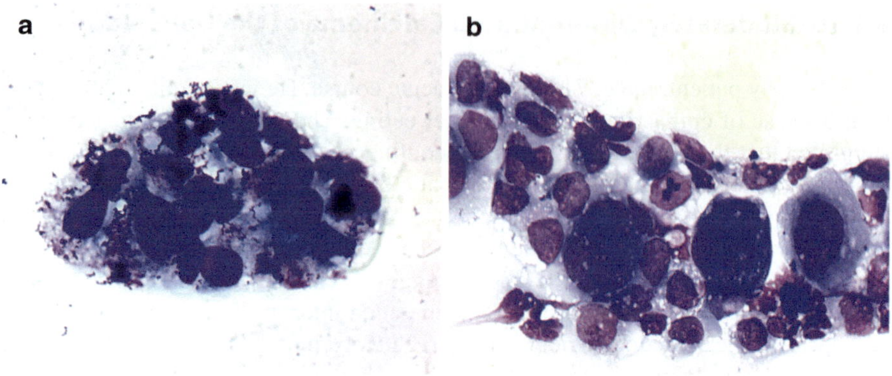

Fig. 3.59 Adenocarcinoma cells in piles and clusters, with disorganized arrangement and obvious heterogeneity (Ritter-Giemsa stain, 1000×)
Moderately differentiated adenocarcinoma shows better cell differentiation, medium cell size, small to medium amount of small acini in the cytoplasm, and morphology similar to moderately nuclear heterogeneous cells and is highly susceptible to missed detection. Under the electron microscope, the nature of the malignant tumor cells is revealed

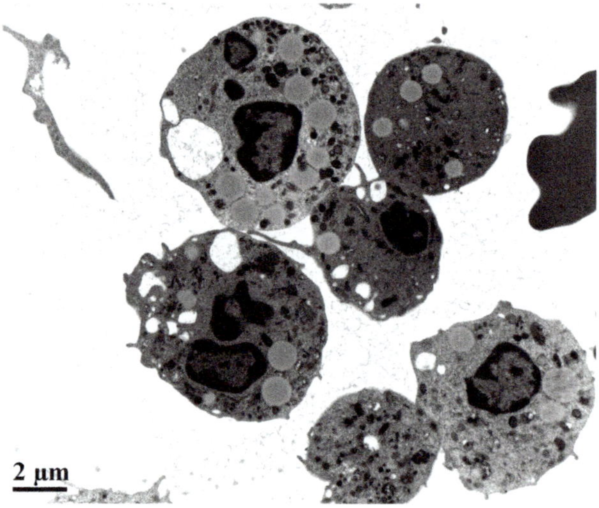

Fig. 3.60 Clusters of moderately differentiated adenocarcinoma cells (TM, 8000×)
Under the electron microscope, the cell size is not large, the nucleus is more mature, the proportion of euchromatin is lower, and the heterochromatin is more, but there are more free ribosomes in the cytoplasm, which makes the cell darker in color, the small acinus is medium in quantity, and other organelles are also abundant. Autophagic vesicles are abundant, and the contents are dissolved when the staining is fixed, so the autophagic vesicles in the figure often appear as white vacuoles, leaving fewer contents. In contrast to the secretory acinus of adenocarcinoma cells, the contents of secretory acini are usually intact, light gray-black, and homogeneous. The two are slightly different

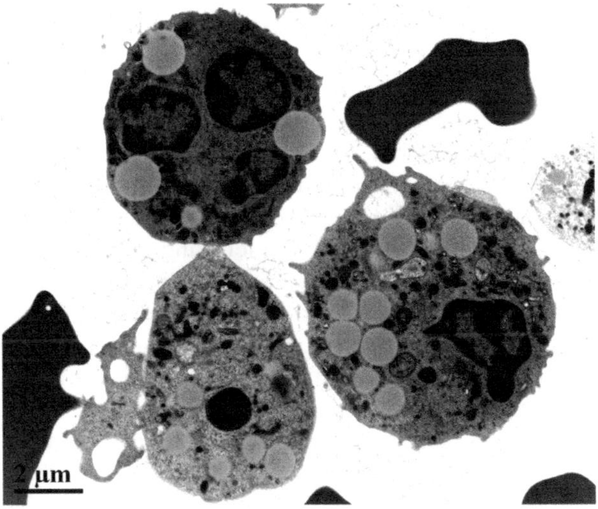

Fig. 3.61 Moderately Differentiated adenocarcinoma cells (TM, 10000×)

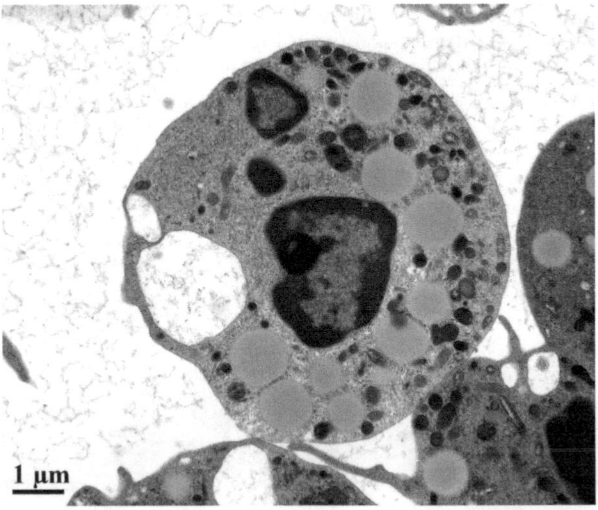

Fig. 3.62 Magnified view of moderately differentiated adenocarcinoma cells (TM, 15,000×) Middle-differentiated adenocarcinoma cells, magnified under electron microscope, are rich in mitochondria and have a large number of free ribosomes, and gray-black acini are densely distributed in the cytoplasm, which can be clearly differentiated from nuclear heterogeneous mesothelial cells

3.1.11 Early Lung Adenocarcinoma in Bronchoalveolar Lavage Fluid

Case: Patient, male, 34 years old, acute course. He presented to the emergency room 1 week ago with fever due to cold, without chills, accompanied by cough and yellow sputum, without chest tightness and shortness of breath, without nausea and vomiting, and without abdominal pain and distension. CT scan of the chest suggests an infectious lesion in the upper lobe of the left lung; please combine the clinical findings with relevant examinations. The patient is admitted to the hospital with the diagnosis of pneumonia. Routine bronchoscopy showed no obvious neoplasm in the bronchoscopic scope. Cytopathologic examination and diagnosis (ciliary bronchoscopic smear of the left lingual branch) showed that no cancer cells were found in the ciliary bronchoscopic liquid-based smear. The patient's bronchoalveolar lavage fluid did not contain many cells, and a small number of scattered or clustered adenocarcinoma cells were detected in the preparation of the smear by cytocentrifugal smear machine. Fluid cytomorphology is shown in Figs. 3.63, 3.64, 3.65, 3.66, and 3.67.

In this patient's liquid-based cytology examination, tumor cells were easily missed due to the small number of nucleated cells, the randomness of the cell wax section, and the high rate of cell loss. By centrifugation and enrichment of the cells, followed by preparation of smears by cytocentrifugation, almost all the cells were enriched on the slide, reducing cell loss and making it easier to detect a small number of tumor cells. Body fluid cytomorphology in early tumor diagnosis highlights the advantages, thanks to the application of enrichment and cell centrifugal smear machine; it improved the detection rate of tumors and other diseases, for the early diagnosis, and early treatment of disease has important clinical significance.

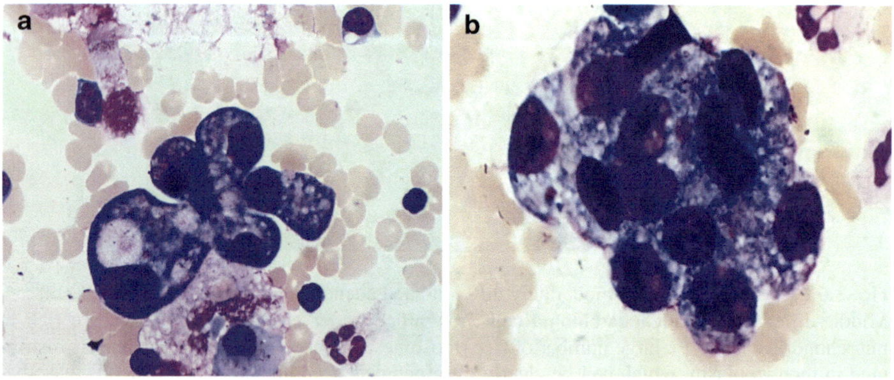

Fig. 3.63 Clustered adenocarcinoma cells (Ritter-Giemsa Stain, 1000×)

Fig. 3.64 Clumps of adenocarcinoma cells, disorganized, loss of polarity, with more complete adenocarcinoma cells visible at the margins (Ritter-Giemsa stain, 1000×)

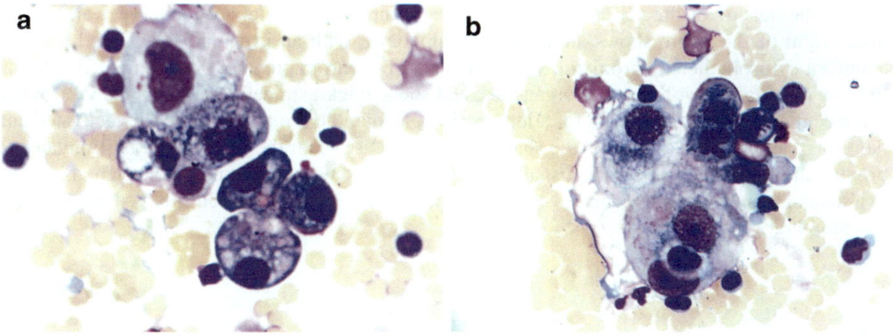

Fig. 3.65 Autophagy and phagocytosis of tumor cells with visible phagosomes (Ritter-Giemsa stain, 1000×)
Tumor cells with disorderly growth and vigorous proliferation, autophagy and phagocytosis are obvious, and abnormal phagocytosis of tumor cells can be seen in Figure B

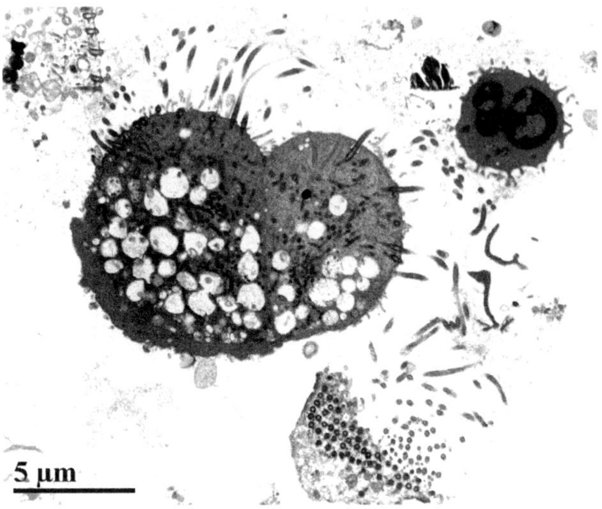

Fig. 3.66 Adenocarcinoma cells and fragmented ciliated columnar epithelial cells (TM, 7000×) The adenocarcinoma cell is covered with cilia from the adjacent fragmented ciliated columnar epithelial cells, the nucleus is squeezed to one side, the nucleus is not cut in this plane, and the malignant features of the cytoplasm are obvious (many free ribosomes, dark coloring, glycogen scattered and appearing in clusters, many acini, abundant mitochondria, and short, dense villi visible on the outside of the cell membrane). Typical lymphocyte nuclei are round or oval, and the upper right of the figure shows an activated lymphocyte with an irregularly depressed nucleus, abundant cytosolic heterochromatin, small amounts of euchromatin, and thick, clumped chromatin; low cytoplasmic volume, few organelles; and short, thick extracellular membrane villi. When the specimen is old or immersed for a long period of time, the lymphocyte is easily activated and deformed, and the first change is the shape of the nucleus; when the lymphocyte is stimulated to produce antibodies, the chromatin usually changes from a state of predominantly heterochromatin to a state of activation with increased euchromatin, and the cell body is often enlarged and has various morphologies

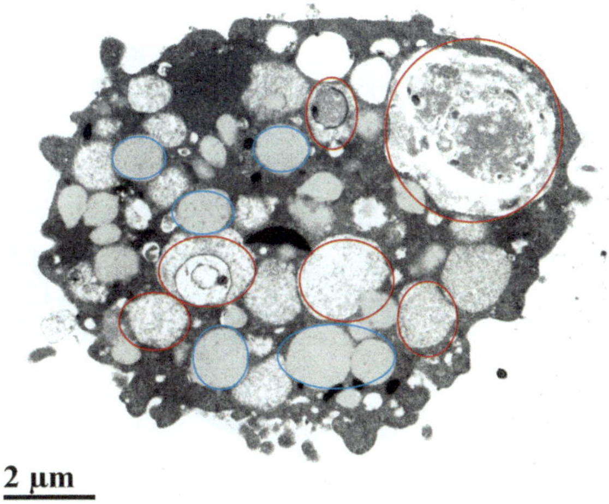

Fig. 3.67 Adenocarcinoma cells (TM, 12000×)
The red circles mark autophagic vesicles or phagocytic vesicles, the contents are partially absorbed or reused, and residual vesicles can be seen, which are common morphological features of tumor cells. Tumor cells rely on active autophagy or phagocytosis to achieve stability of the internal environment and reuse of materials to maintain their rapid proliferation and development. The blue circle marks the secretory vesicles, which are uniform and gray-black. Adenocarcinoma cells have abundant secretory vesicles, which are related to their cellular origin. Tumor cells of glandular origin often have abnormal secretory functions, and some of them can also secrete hormone-like substances, leading to tumor-associated endocrine abnormality syndromes

3.1.12 Lung Adenocarcinoma with Multiple Metastases

Case: Male, 66 years old, chronic disease with acute exacerbation, diagnosed with lung malignant tumor (adenocarcinoma) 6 years ago and treated with various chemotherapy regimens and targeted therapies, with recurrent exacerbations, during which the patient suffered from pleural effusion several times and was given a thoracentesis and drainage. The patient was admitted to the hospital for treatment of "respiratory failure, lung malignant tumor with metastasis." Gastroscopy showed "gastric wall nodule" with small pieces of inflammatory fibrous tissue and granulation tissue, covered with inflammatory exudate, which was consistent with pathological changes of gastrointestinal perforation. Chest CT report showed narrowing of the right lung bronchus, partially opacified, large pleural effusion with atelectasis on the right side; a few patchy shadows on the left lung; no enlargement of the left hilar; the main trachea and the left bronchus were clear; enlarged mediastinal lymph nodes; a small amount of pleural effusion on the left side with incomplete distension of the neighboring lung tissues; nodular thickening of the pleural membrane on the right side; subcutaneous edema of the chest wall on both sides; dilatation of the esophagus; and an accompanying peritoneal effusion. The cytomorphology of body fluids is shown in Figs. 3.68, 3.69, 3.70, 3.71, 3.72, 3.73, 3.74, 3.75, and 3.76.

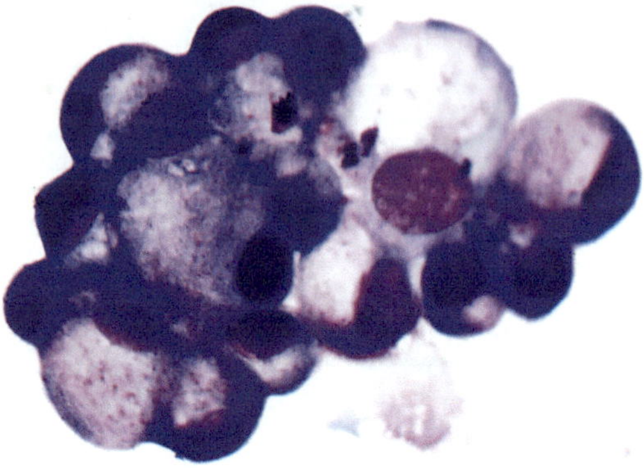

Fig. 3.68 Clustered adenocarcinoma cells (Ritter-Giemsa stain, 1000×)

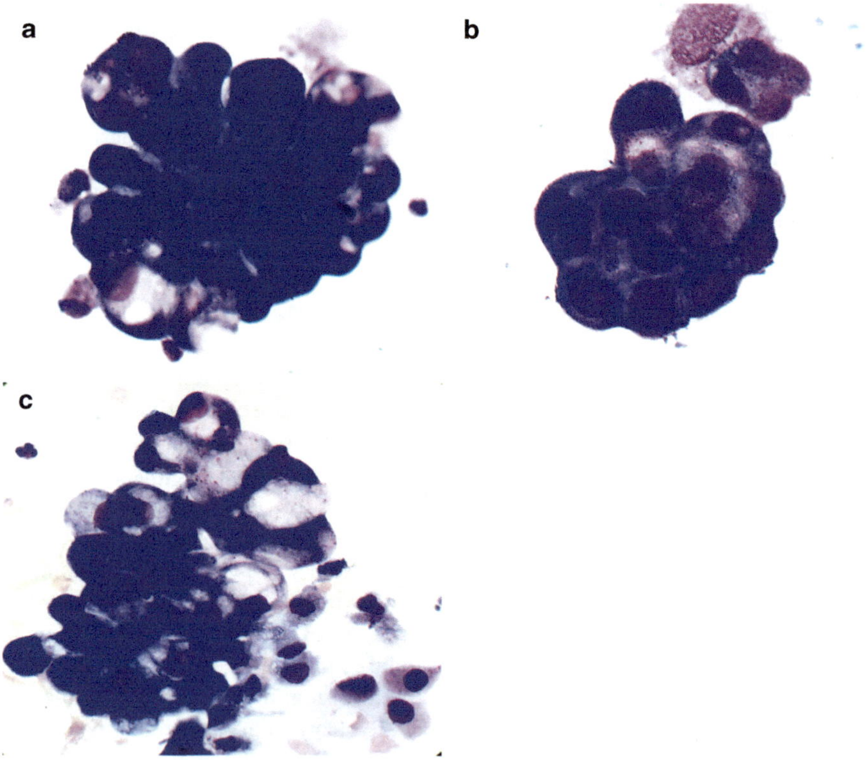

Fig. 3.69 Arranged and disorganized cancer nests (Ritter-Giemsa stain, 1000×)
Individual tumor cells are small in size, piled up in clusters and arranged in a typical glandular cavity-like cancer nest, which has the characteristics of tumor cell arrangement: disorganized polarity, different sizes, glandular cavity-like, and unclear hierarchy. The purple dense particles under light microscope were black autophagic vesicles under electron microscope

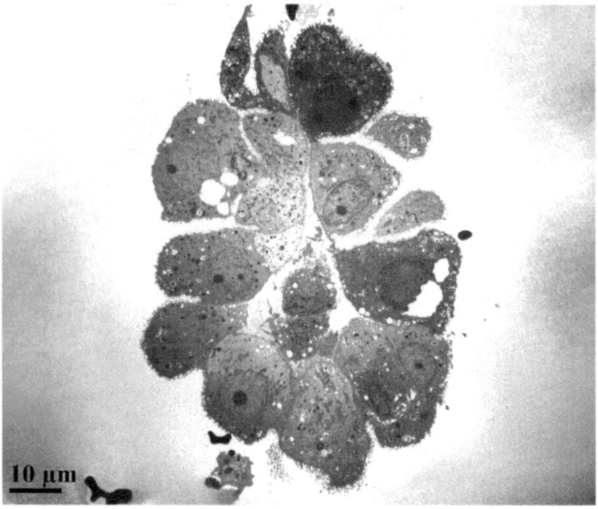

Fig. 3.70 Intact cancer nest with disorganized arrangement and obvious cellular heterogeneity (TM, 1500×)

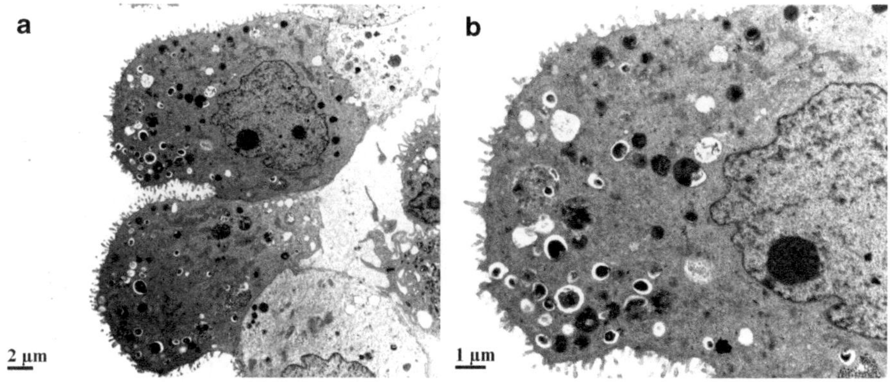

Fig. 3.71 Magnified electron micrograph of a localized cancer nest (A. TM, ×5000; B. TM, 12,000×)
Purple-red granules under light microscopy and autophagic or phagocytic vesicles under electron microscopy. The extracellular microvilli were dense, the nuclei were malformed, euchromatin was dominant, the nucleoli were large, the cytoplasm was rich in small free ribosomes, the cell body was black, the glycogen was distributed in clusters, and the intercellular boundary was clear

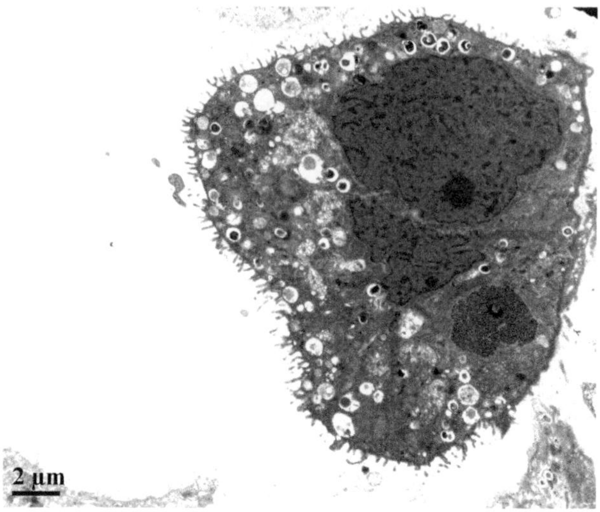

Fig. 3.72 Single adenocarcinoma cells (TM, 7000×)
Tumor cells are highly malignant, with malformed nuclei, obvious nucleoli, abundant phagocytic vesicles in the cytoplasm, high number of free ribosomes, abundant glycogen aggregated into large clumps, darkly colored cell body, and abundant mitochondria. Microvilli on the cell membrane are abundant and dense, which are tools to assist tumor cell adhesion, implantation and metastasis

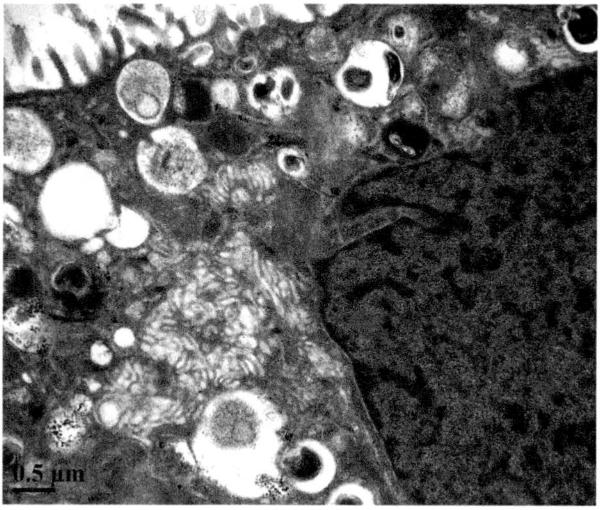

Fig. 3.73 Localized magnified electron micrograph of a single adenocarcinoma cell (TM, 25,000×)
Electron micrographs reveal well-developed endoplasmic reticulum, various autophagic and phagocytic vesicles, and abundant glycogen deposits

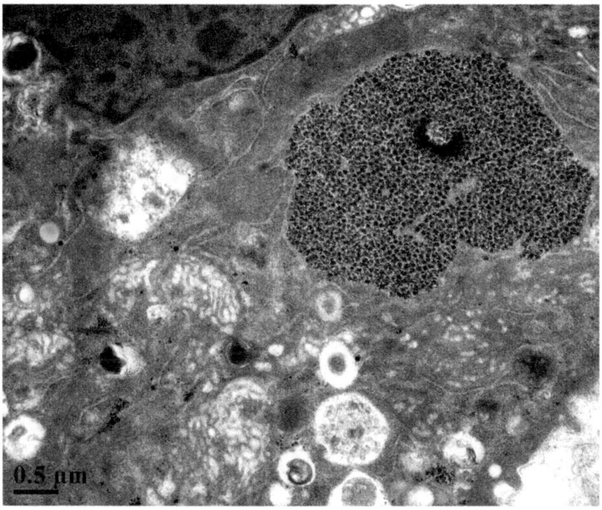

Fig. 3.74 Localized magnified electron micrographs of adenocarcinoma cells (TM, 25,000×)
Large clusters of glycogen α particles are seen, and the endoplasmic reticulum system is well developed. Some of the endoplasmic reticulum gaps are widened and the membrane integrity is weak, which suggests apoptotic tendency

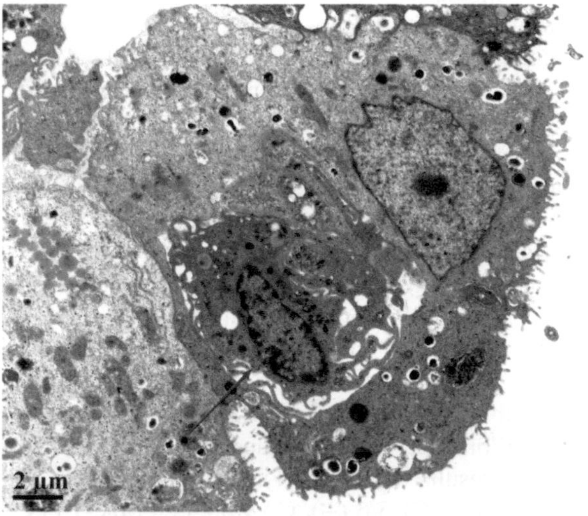

Fig. 3.75 Tumor cells containing phagosomes (TM, 7000×)
Tumor cells engulfed intact a macrophage (arrow) whose nucleus contained more heterochromatin, was irregular, had no nucleolus, had phagocytic vesicles formed by secondary lysosomes, and had abundant, elongated, folded and twisted pseudopods on the cytosolic membrane, unlike the tumor cells' fine, short microvilli

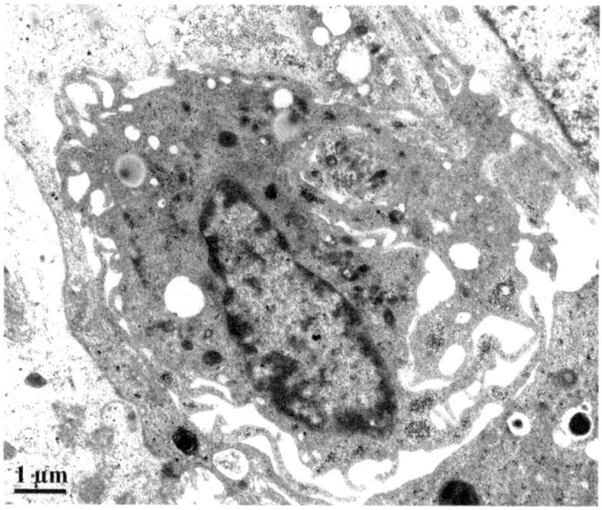

Fig. 3.76 Phagocytized macrophages (TM, 15,000×)
Clusters of tumor cells can often be seen engulfing each other with other cells, possibly as a game between the body's defense system and the tumor, and perhaps closely related to tumorigenesis, growth and metastasis

3.1.13 Moderately Differentiated Adenocarcinoma of the Bile Ducts of the Hepatoportal Region

Patient: Female, 72 years old, with chronic disease. Seven months ago, due to fever with scleral yellowing, she was diagnosed with hepatoportal cholangiocarcinoma with obstructive jaundice and underwent "resection of gallbladder and extrahepatic bile ducts, dissection of lymph nodes of hepato-duodenal ligament, hepatoportal cholangioplasty, and hepatic portal jejunostomy" under general anesthesia. During the operation, an irregular hard mass was palpable at the confluence of the right anterior lobe hepatic duct and the left hepatic duct, and the lower edge was 2 cm away from the pancreas. The bifurcation of the portal vein and the left and right branches of the portal vein were infiltrated by the tumor, with the circumference of the invasion being more than 1/2 of the circumference, and the right hepatic artery was surrounded by the tumor. Postoperative pathology report showed (1) extrahepatic biliary adenocarcinoma, 2.5 cm × 1.2 cm, moderately differentiated, infiltrating the whole bile duct wall and invading the nerves; (2) no metastasis in the regional lymph nodes and parabiliary lymph nodes (0/1, 0/1); (3) chronic calculous cholecystitis of the gallbladder, accompanied by gallbladder adenomyosis. Immunohistochemistry results were CK7 (+), CEA (+), CK19 (+), Villin (+), Hep-1 (−), HBsAg (−), CK20 (−), CK5/6 (−), CAM5.2 (+), Muc-1 (+), and HER2 (−). Genetic testing revealed a KRAS mutation. Magnetic resonance cholangiopancreatography (MRCP) report suggested that after surgery for hepatoportal cholangiocarcinoma, there was a possibility of tumor recurrence in the left lobe of the liver and the hepatoportal area, soft tissue shadow in the retroperitoneum, and a small amount of ascites. The patient had recurrent fever and started to have poor appetite 2 weeks ago, accompanied by nausea, abdominal distension, fatigue, and abdominal pain at times, which was tolerable, with white mucous sputum that could not be easily coughed up. There was no chest tightness or palpitation, and she did not stop defecating or passing gas, so she was admitted to the hospital as a case of hepatoportal cholangiocarcinoma. The cytomorphological examination of ascites was shown in Figs. 3.77, 3.78, 3.79, 3.80, and 3.81.

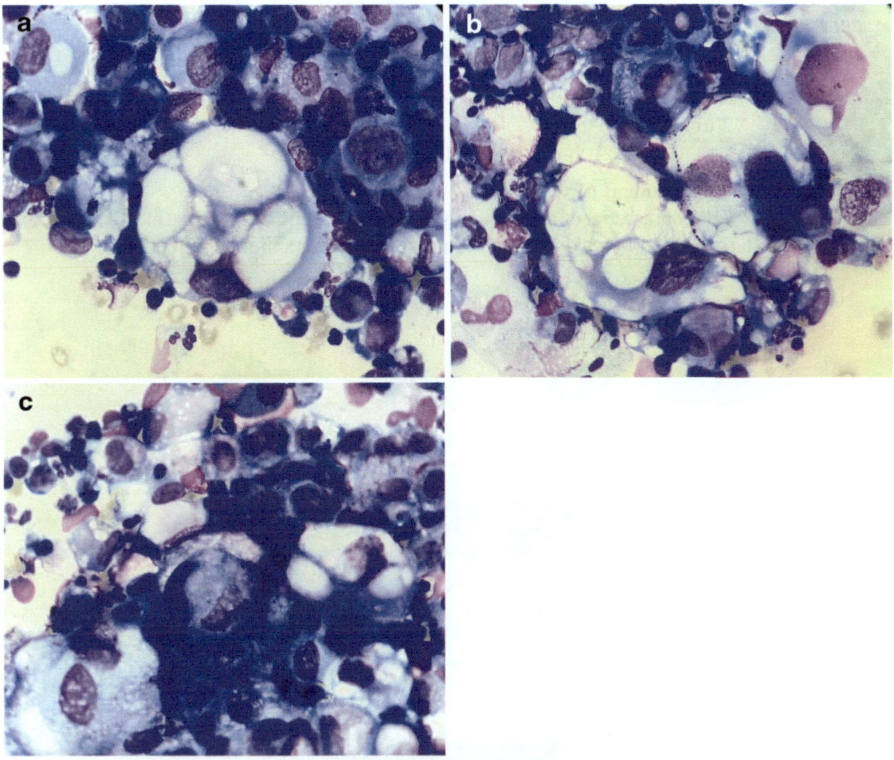

Fig. 3.77 Typical adenocarcinoma cells and nests (Ritter-Giemsa stain, 1000×)
Phagocytosed and chimeric tumor cells can be seen within the tumor cell mass, which are disorganized, unpolarized and nonhierarchical. The tumor cells in the cancer nests have different degrees of differentiation, and those with a higher degree of differentiation are larger in size and have obvious acinus, while those with a lower degree of differentiation are small in size, with a large nucleocytoplasmic ratio and small and inconspicuous acinus

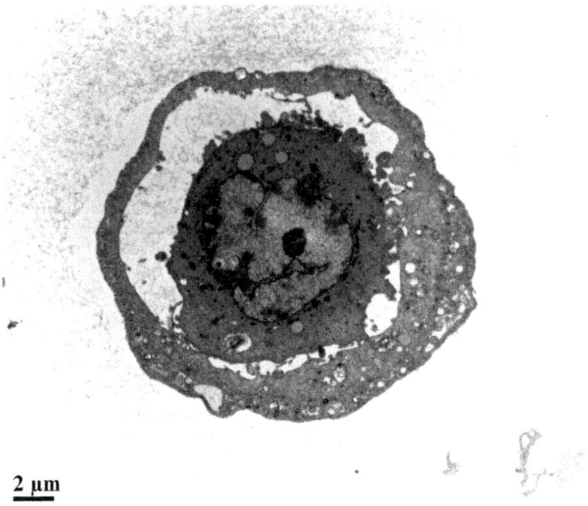

Fig. 3.78 Tumor cell chimeras (TM, 6000×)
Tumor cells are well characterized, and phagocytosed tumor cells are poorly differentiated with distinct nuclei, large nucleo-cytoplasmic ratios, and fewer markers of maturation

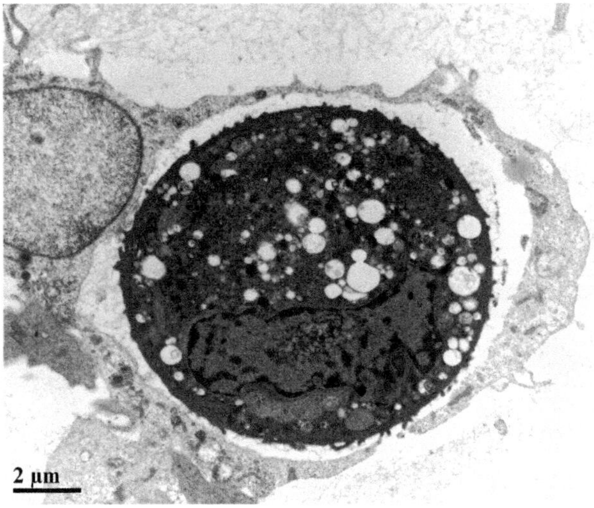

Fig. 3.79 Tumor cells encapsulated by apoptotic tumor cells (TM, 10,000×)
The apoptotic tumor cell in the figure wraps around another tumor cell. Cells tend to apoptose themselves after engulfing larger cells

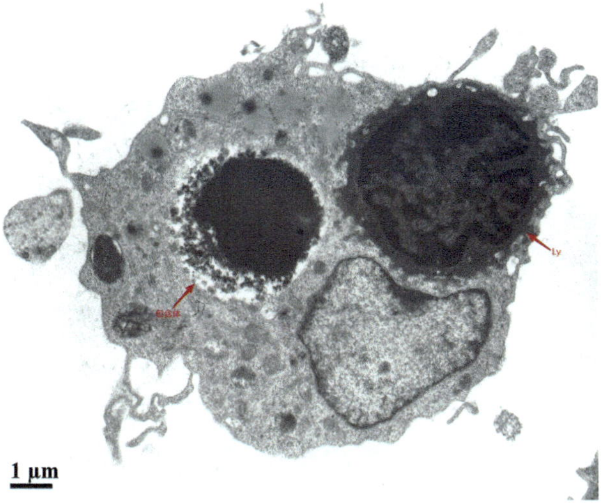

Fig. 3.80 Tumor cell phagosomes (TM, 12,000×)
Tumor cell phagosomes contain large inclusion bodies and a lymphocyte. Tumor cell nuclei are predominantly euchromatin with abundant acini; lymphocytes are predominantly aggregated heterochromatin with less plasma; inclusion bodies are poorly structured, dense and homogeneous

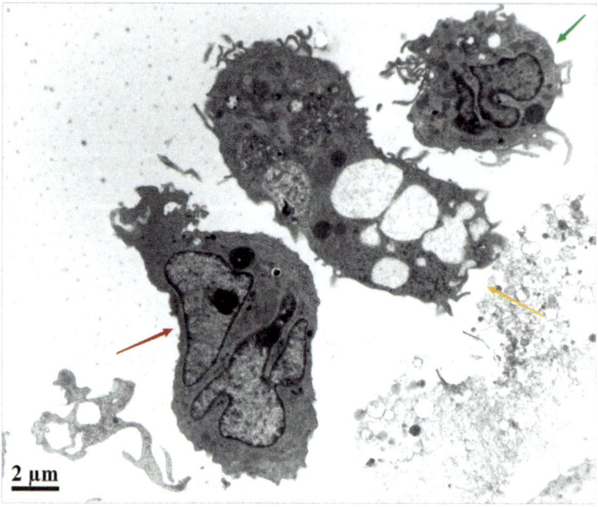

Fig. 3.81 Adenocarcinoma cells (TM, 10000×)
The tumor cells in the figure have obvious nuclear malformations and belong to middle differentiated adenocarcinoma, some cells are more naïve, while some cells are more mature. The adenocarcinoma cells with obvious nuclear malformations shown by the red arrows are more naïve, with a large and obvious nucleolus, euchromatin dominated, and less heterochromatin mass; the nucleoplasma ratio is large, cytoplasm is small, and the development of acinus is not obvious, which is obviously more naïve than that of the intermediate tumor cells. The cell shown by the yellow arrow is a mature adenocarcinoma cell. Although the nucleus is not cut in this plane, the large and dense acini in the cytoplasm indicate that it is relatively mature, with a lot of free ribosomes, rich in glycogen, and grayish-black in color. The cell shown in the green arrow is an activated macrophage, the cell body is not as large as the tumor cell, the pseudopods are elongated, the nucleus is folded and twisted, the chromatin is more mature, the heterochromatin is more than that of the tumor cell, there is no nucleolus, the cytoplasm is rich in phagocytosis vesicles and lysosomes, the cell color is darker because of the phagocytosis of more stuffs, and the color of the cell is deeper than that of the un-activated macrophage

3.1.14 Rectal Adenocarcinoma with Multiple Systemic Metastases

Case: Patient, male, 60 years old, with chronic disease. Three years ago, he underwent colonoscopy, and the report suggested rectal cancer, polypectomy of the colon. Pathologic findings suggested "rectal mass biopsy" adenocarcinoma. After admission to the Department of Anorectal Surgery, he underwent abdominal CT, rectal magnetic resonance and whole-body PET-CT, which indicated rectal cancer with multiple metastases to the liver and lymph nodes, and was staged as cT4bN2M1b. After multidisciplinary consultation and discussion, mFOLFOX6 regimen was administered, followed by mFOLFXOX+bevacizumab regimen for six times, with regular review and evaluation of the therapeutic efficacy of PR. One year and 9 months ago, the liver magnetic resonance report suggested multiple nodules in the liver, considering metastatic foci, and the foci were enlarged compared with the previous film (CT report). He was admitted to the hospital with the diagnosis of rectal cancer. The cytomorphologic examination of pleural fluid is shown in Fig. 3.82, 3.83, 3.84, 3.85, 3.86, 3.87, and 3.88.

The tumor cell growth doctrine suggests that the doubling time of a tumor cell refers to the time it takes for the cell to divide and multiply into two daughter cells. Most malignant cells do not have a faster multiplication time than normal cells, so the rapid growth of malignant tumors may not be primarily due to a shortening of the multiplication time of tumor cells but rather to an increase in the proportion of cells in the proliferative state in the tumor cell population. The growth fraction of tumor cells refers to the proportion of proliferating cells in the tumor cell

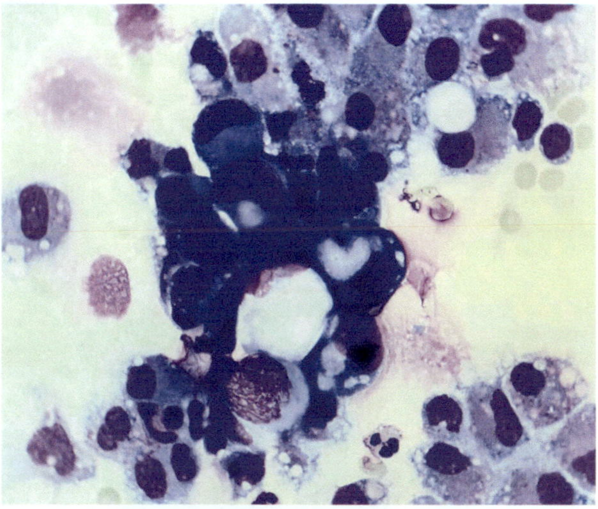

Fig. 3.82 Carcinoma nest (Ritter-Giemsa stain, 1000×), disturbed polarity of cell arrangement

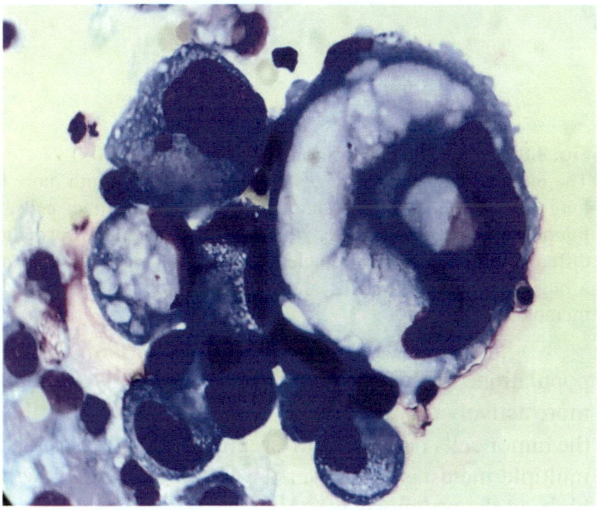

Fig. 3.83 Adenocarcinoma Cells (Ritter-Giemsa Stain, 1000×) Adenocarcinoma cells with typical morphology, huge cell body, large nucleus, large nucleo-cytoplasmic ratio, three-dimensional and abundant acinus, cloudy, nucleus squeezed to the edge of the cell, and pronounced nuclear deformity

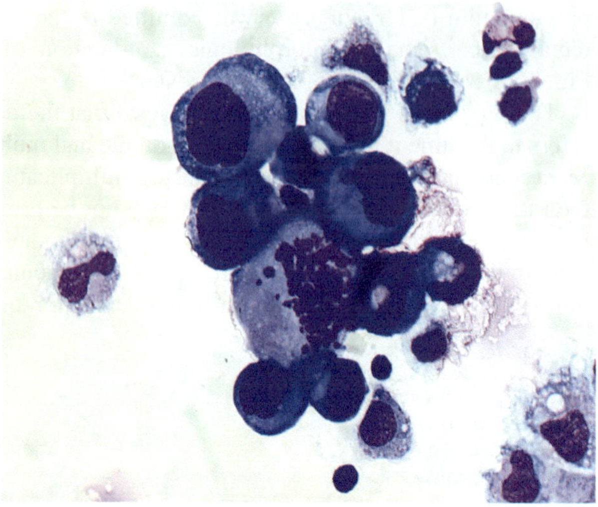

Fig. 3.84 Mitotic figure of tumor cells (Ritter-Giemsa stain, 1000×) Aneuploidy is common, irregularly arranged, and tumor cells are of different sizes

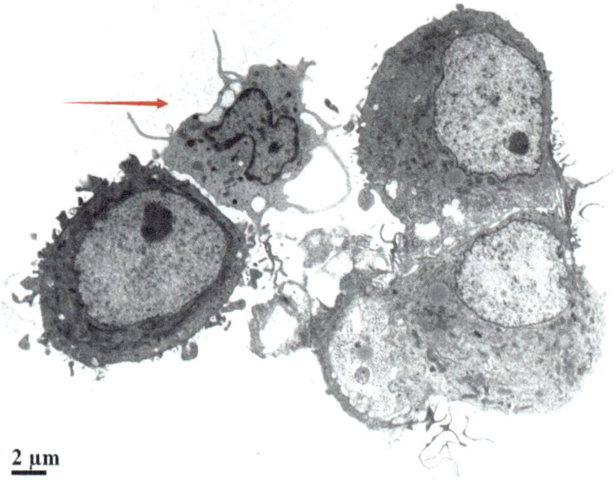

Fig. 3.85 Clustered adenocarcinoma cells (TM, 5000×)
The red arrows in the figure point to macrophages, with more heterochromatin, irregular nuclei, long pseudopods, and fewer numbers; the rest are tumor cells, the tumor cells on the left of the figure are poorly differentiated; the tumor cells on the lower right of the figure are relatively highly differentiated, with a small nucleoplasma ratio, inconspicuous nucleoli, abundant cytoplasm, and a huge acinus, but the cells are in the apoptotic phase, with loose chromatin and unclear cell membranes

population. Proliferating cells are constantly dividing and reproducing, and the more actively they divide and reproduce, the higher the growth fraction. In this case, the tumor cells had a low differentiation degree; the disease progressed rapidly, with multiple metastases all over the body; and the growth fraction of tumor cells was high, so the prognosis was poor.

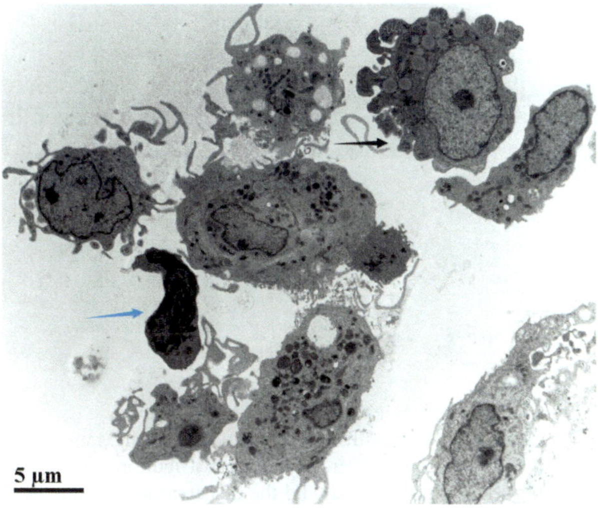

Fig. 3.86 Electron micrograph of a tumor cell (TM, 4000×)
The black arrow in the figure is a tumor cell (naïve nucleus, obvious nucleolus, abundant euchromatin, cytoplasmic glycogen appearing in clusters, small and many acini, abundant mitochondria, abundant endoplasmic reticulum); the blue arrow is a lymphocyte (large nucleoplasmic ratio, nucleus with predominantly heterochromatin, strong smearing, little cytoplasm, fewer organelles); the rest are macrophages, light-colored, lysosome-rich, irregular nucleus, more or less cytoplasm, pseudopods long and folded and curled

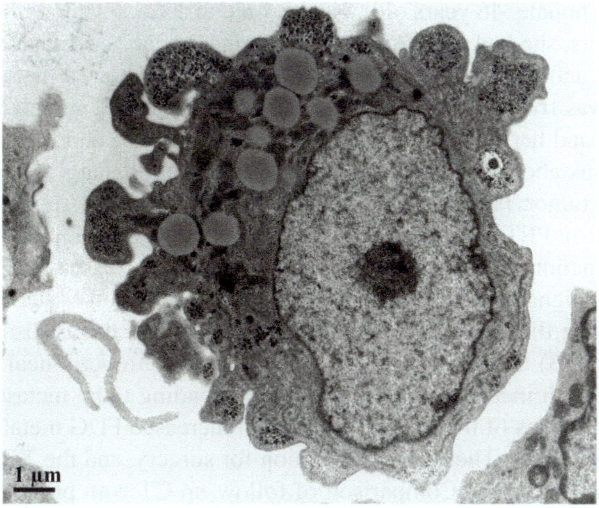

Fig. 3.87 Magnified electron micrograph of tumor cells (TM, 5000×)
The tumor cells contained abundant glycogen, clustered within the cytoplasm, along with numerous mitochondria and variably sized, small acini. The endoplasmic reticulum was arranged in concentric, perinuclear layers. The nuclei were predominantly composed of euchromatin and featured large, "bull's-eye" nucleoli. This case showed a high proportion of immature tumor cells, all of which were small, with a high nuclear-to-cytoplasmic ratio and prominent nucleoli

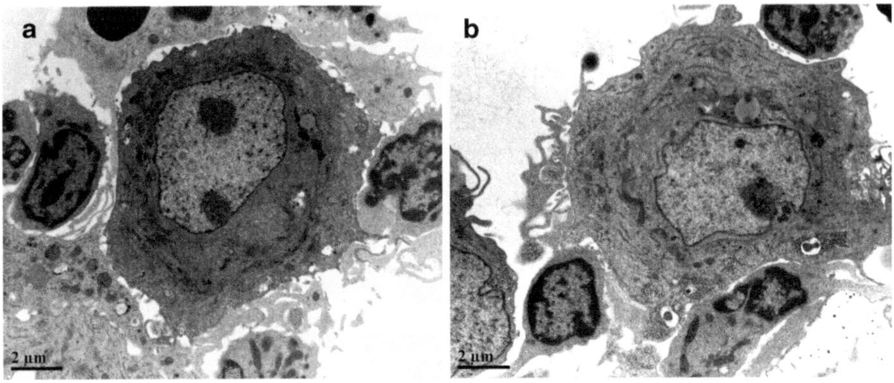

Fig. 3.88 Less differentiated adenocarcinoma cells (TM , 10 000×)
The tumor cells were all small in size, with large nucleo-cytoplasmic ratio and large nucleoli like bull's eye, suggesting that the tumor cells were naïve; the distribution of glycogen in clusters in the cytoplasm, the high number of mitochondria, and the distribution of endoplasmic reticulum cascading around the nucleus suggested that the tumor cells were growing and proliferating rapidly. The patient had a poor prognosis and passed away not long after sampling, which fully demonstrated that patients with high growth fraction of tumor cells metastasized and progressed more quickly

3.1.15 Rectal Adenocarcinoma with Multiple Metastases

Case: Patient, female, 46 years old, with chronic disease. Five months ago, without obvious triggers, she had persistent pain in perianal area, accompanied by abdominal distension and constipation, and her tumor markers showed that carcinoembryonic antigen was 16.6 ng/ml, glycoantigen 125: 1418.6 U/ml, and glycoantigen 199: 3488.5 U/ml, and her colonoscopy report showed that she had rectal mass (firstly, she should think about the tumor). Enteroscopy ES report suggested rectal occupation, consider tumor. Pathologic biopsy result of colonoscopy suggested adenocarcinoma (rectum). PET-CT report suggested (1) obvious thickening of the intestinal wall in the junction area of rectosigmoid colon with increased FDG metabolism, considering malignant lesions; (2) mass and strip-shaped soft tissue shadow next to the left common iliac blood vessel with increased FDG metabolism, considering metastasis; and (3) enlarged perirectal fat space and retroperitoneal lymph nodes, part of which with increased FDG metabolism, tending to be metastatic. Multiple nodules in both lobes of the thyroid gland with increased FDG metabolism; malignant lesions are likely. There is no indication for surgery, and the XELOX regimen is used for chemotherapy. Comparison of follow-up CT with previous CT showed (1) scattered multiple small nodules in both lungs, metastatic possibility, roughly similar to the previous one; (2) multiple slightly low-density nodules in the liver, enlarged and increased compared to the previous one, metastatic consideration; (3) hemangioma in the S6 segment of the liver, similar to the previous one, and gallbladder stones; and (4) multiple small lymph nodes in the retroperitoneum, similar to the previous one. Attachment: nodules in both lobes of the thyroid gland, please combine with other investigations. This suggests progression of the disease. A large number of tumor cells can be seen in the ascites. The morphology of body fluid cells is shown in Figs. 3.89, 3.90, 3.91, 3.92, 3.93, 3.94, and 3.95.

Factors affecting apoptosis include inhibitory and inducing factors: the former include growth factors, cellular matrix, steroid hormones and certain viral proteins; the latter include growth factor deficiency, glucocorticoids, free radicals and ionizing radiation. There are dozens of genes involved in apoptosis, and genes such as c-myc may have a bi-directional regulatory role. When growth factors are sufficient, they promote cell proliferation, and when they are deficient, they cause apoptosis.

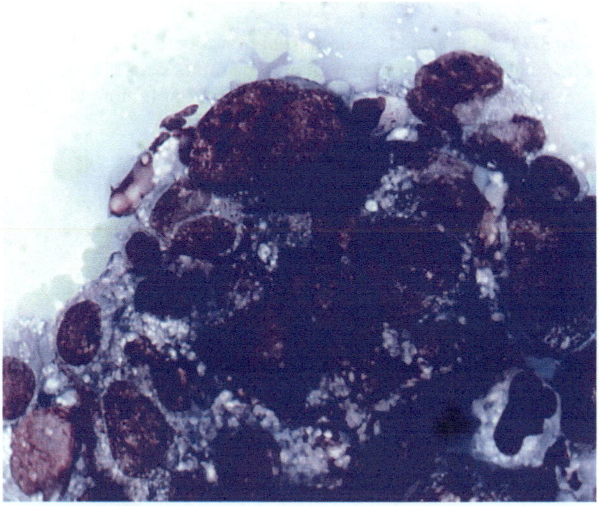

Fig. 3.89 Clustered adenocarcinoma cells (Ritter-Giemsa stain, 1000×)
The adenocarcinoma cells are distributed in clusters, the intercellular space is not obvious, the cells at the edge of the cancer nest are clearer, and the characteristics of the tumor cells can be observed. The nuclei are obviously malformed; the cells have less cytoplasm, more nucleoli, and more acini

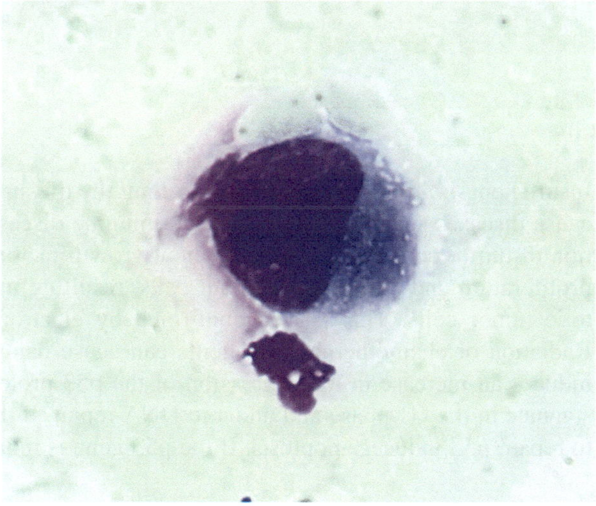

Fig. 3.90 Naked nucleus-like tumor cells (Ritter-Giemsa stain, 1000×)
In this case, more tumor cells with naked nuclei were seen, with little or no cytoplasm, a large nucleo-cytoplasmic ratio, obvious nucleoli, varying cell size, obvious nuclear aberrations, and tiny acinus in the cytoplasm. The cells were broken, suggesting that the cells were in a pre-apoptotic or apoptotic state

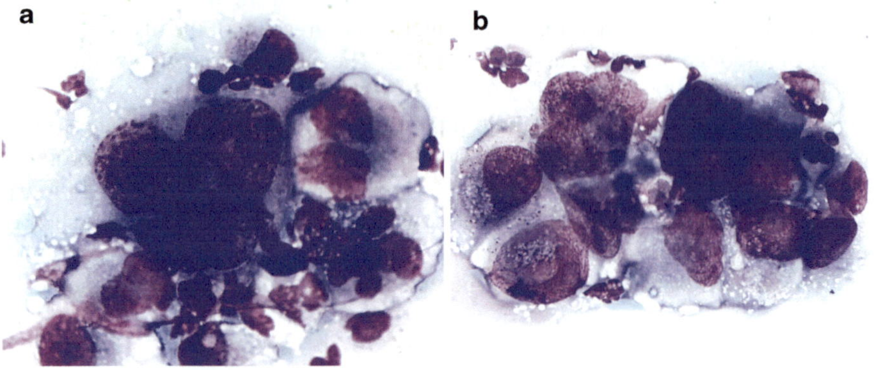

Fig. 3.91 Malformed, disorganized tumor cells (Ritter-Giemsa stain, 1000×)
The polarity of the arrangement of these tumor cells is lost, with no hierarchy or direction, and they are stacked randomly, reflecting the disorder of tumor cell growth

Fig. 3.92 Naked nucleus-like tumor cells (TM, 15,000×)
Naked nucleus-like cells, very little cytoplasm visible on electron microscopy, few organelles in the cytoplasm, only one phagocytic vesicle; loose chromatin in the nucleus, sense of apoptosis

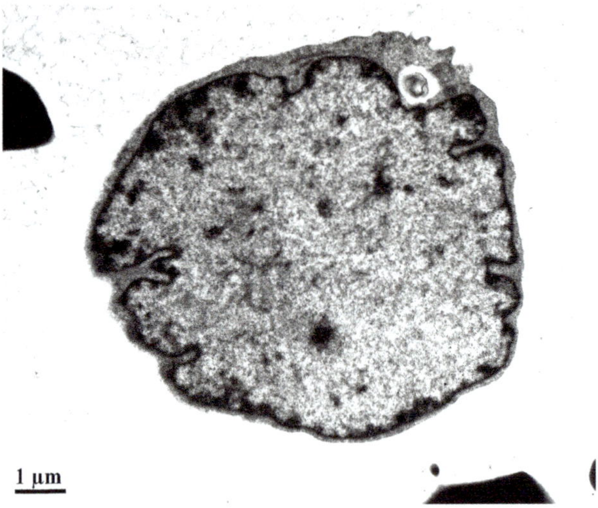

Insufficient or absent apoptosis can prolong the lifespan of the cells concerned and cause diseases such as tumors and autoimmune diseases. The excessive proliferation of tumor cells results in insufficient growth factors to sustain the high rate of proliferation and development of all cells, resulting in a high prevalence of apoptotic tumor cells. This is also confirmed by electron microscopic observations. Radiation or chemotherapeutic agents can cause damage to cellular DNA, which induces an increase in the expression of the p53 protein, which causes the cell to stagnate in the G1 phase and undergo DNA repair. If the DNA damage is too great to repair, p53 induces apoptosis. If the p53 gene is mutated or lost and apoptosis is

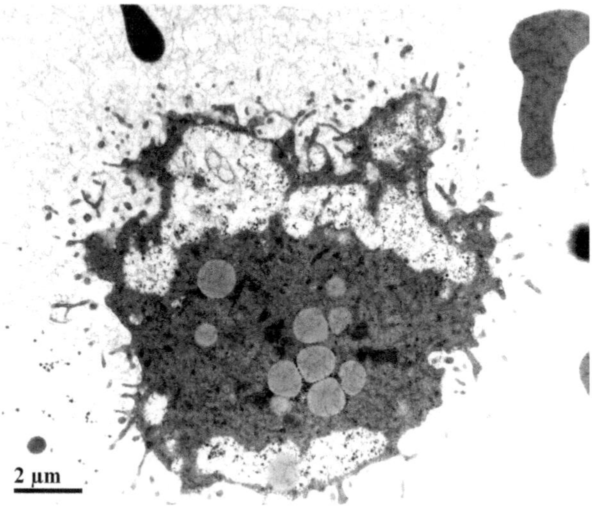

Fig. 3.93 Tumor cells in the disintegrated state (TM, 10,000×)
Tumor cells in a disintegrated state, no nucleus visible, some acini in the cytoplasm fused into one, a few acini and organelles in the center of the cell, cell membrane incomplete

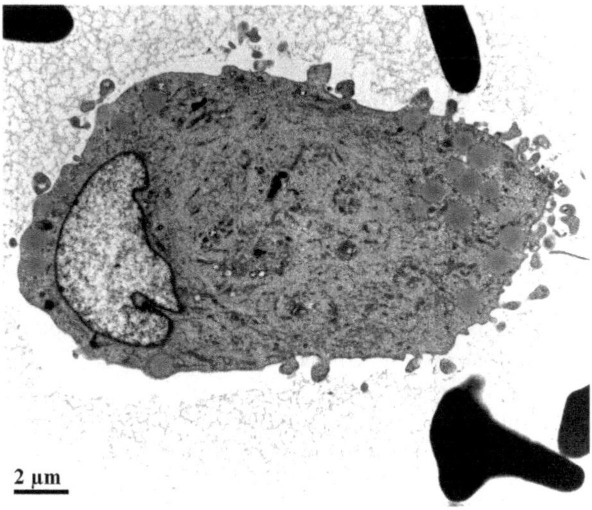

Fig. 3.94 Tumor Cells (TM, 8000×)
Tumor cell body is intact, but the nucleus shows obvious signs of apoptosis, and the chromatin is loose and Incomplete structured. The organelles in the cytoplasm are Incomplete structured, broken, and lysed easily seen

not induced, the cell survives and eventually the DNA-damaged cell undergoes malignant transformation and a tumor forms. If the telomerase enzyme remains active (telomeres shorten by 50–200 nucleotides with each mitotic division until the cell is senescent and no longer dividing, and the telomerase enzyme can reverse transcriptionally lengthen the shortened telomeres), the cell is immortalized, as in the case of HeLa cells. The patient's tumor cells were mostly apoptotic and did not belong to the immortalized series of tumor cells. The growth fraction of his tumor was high, with obvious metastases in various parts of the body and a poor prognosis.

3.1.16 Invasive Lobular Carcinoma (Adenocarcinoma) Metastasis of the Breast

Case: Patient, female, 64 years old, with chronic disease. In 1997, she was diagnosed with left breast cancer due to left breast enlargement and underwent radical surgery for left breast cancer, after which she did not undergo radiotherapy or endocrine therapy, and no obvious tumor recurrence or metastasis was found in her regular check-ups. In August 2016, the patient touched a lump on the left chest wall near the parietal sternum by herself, but she did not pay attention to it. In February 2017, tumor markers were found to be elevated, and the breast ultrasound report suggested postoperative left breast and substantial left anterior thoracic occupation (invasion of ribs). Pathology of left chest wall mass showed: "left chest wall puncture" invasive carcinoma. Immunohistochemical staining showed P53 (−), Calponin (−), GCDFP15 (−), P120 (membrane +), P63 (−), AR (+), CK5/6 (−), ER (+95%, strong), HER2 (1+), Ki-67 (+20%), and PR (+15%, moderate). 8 March 2017 Whole body bone imaging. On 8 March 2017, the whole body bone imaging report suggested punctate bone metabolism active foci in the sternum, combined with CT, bone metastasis was not excluded. On 11 March 2017, the liver and gallbladder enhancement magnetic resonance report suggested multiple nodular shadows in the hepatoportal area, combined with the history of the disease, lymph node metastasis was considered. After exclusion of contraindications, first-line chemotherapy with TX (paclitaxel + capecitabine) regimen was administered for six cycles, followed by palliative radiotherapy (DT 5000 cGy/25f) for a subcutaneous mass on the left chest wall. After that, the disease was stable. On 3 September 2018, the ultrasound results of breast and axillary lymph nodes showed a hypoechoic area in the left parasternal intercostal space below the clavicle, with a general range of 29 mm × 19 mm × 24 mm, and several hypoechoic nodules in the right axilla, with a larger size of 14 mm × 6 mm, and metastatic foci were considered as a possible cause of the disease. Ultrasound results of supraclavicular and cervical lymph nodes indicated that multiple lymph nodes were detected in the bilateral neck and supraclavicular region, with the larger ones being 34 mm × 19 mm (left cervical region IV) and 13 mm × 8 mm (right cervical region IV), which was considered to be metastatic foci. On 12 September 2018, the routine pathology result showed "puncture biopsy of left supraclavicular mass" malignant tumor, considering lymph node metastatic cancer. Immunohistochemical staining results: CAM5.2 (+), CK (Pan) (+), CD38 (−), CD79a (−), Mum-1 (−), ER (95%3+), PR (30%2+), HER-2 (2+), Ki-67 (28%+), P120 (plasma +), AR (+), GATA3 (+), E-cadherin (−), P63 (−), CK5/6 (−), mammaglobin (+), CgA (−), SYN (−), CD56 (−), CD138 (+), SOX10 (−), GCDFP-15 (−), and actin (HHF35) (−). The combination of immunohistochemical markers was consistent with metastatic invasive lobular carcinoma of the breast. Results of fusion gene testing: HER2 gene was not amplified (negative).

Diagnosis: ① endocrine therapy for malignant tumor; ② malignant tumor of left breast (postoperative, invasive lobular carcinoma, stage IV), secondary malignant tumor of several lymph nodes (neck, supraclavicular, hepatic hilum, axilla, after local radiotherapy), secondary malignant tumor of chest wall (after local radiotherapy), secondary malignant tumor of skeleton (after radiotherapy of right sciatic bone); ③ hypothyroidism. The patient had extensive metastasis of breast cancer, and the cytomorphological examination of pleural fluid is shown in Fig. 3.95, 3.96, 3.97, 3.98, 3.99, 3.100, 3.101, 3.102, 3.103, and 3.104.

This patient has invasive lobular carcinoma of the breast, which is a type of invasive non-specific carcinoma. This type is generally less differentiated and has a poorer prognosis and is one of the most common types of breast cancer. The presence of distant bone metastasis indicates that it is caused by hematogenous metastasis. As the mammary gland has secretion function, there are often a large number of secretory vesicles in the cytoplasm of its tumor cells. Some tumor cells of organ origin with secretory function can secrete hormones and participate in the neuroendocrine regulation of the body, showing complex symptoms and leading to paraneoplastic syndrome. The characteristics of such cells can be observed under the electron microscope.

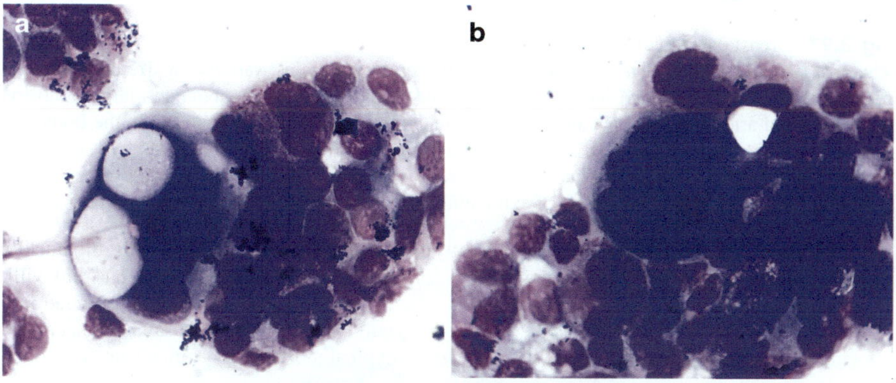

Fig. 3.95 Clustered breast cancer cells (Ritter-Giemsa Stain, 1000×)

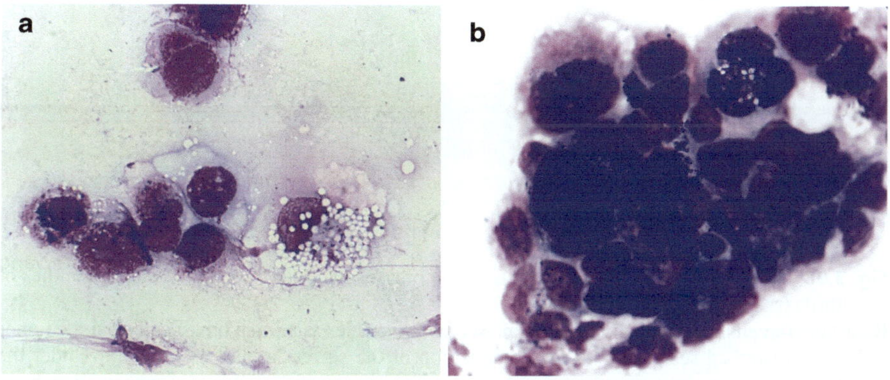

Fig. 3.96 Tumor cells, (**a**) scattered; (**b**) clustered (Ritter-Giemsa stain, 1000×)

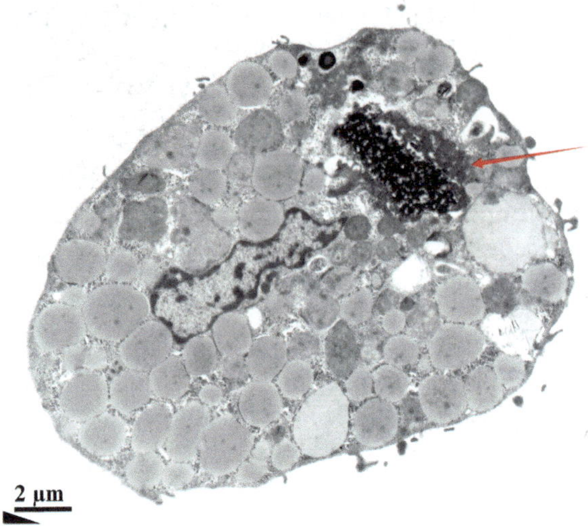

Fig. 3.97 Breast cancer cells (TM, 8000×)
This is a typical breast cancer cell, with abundant secretory vesicles, variable in size and homogeneous; the nucleus is extruded into a flattened shape and euchromatin is abundant; large inclusion bodies formed after phagocytosis are visible in the cytoplasm, containing material in the body that has been digested to the point of structural ambiguity (red arrows); densely populated free ribosomes and glycogen are visible in the crevices of the secretory vesicles in the cytoplasm, with some of them appearing in clusters; a small number of extracellular microvilli are visible and are relatively short

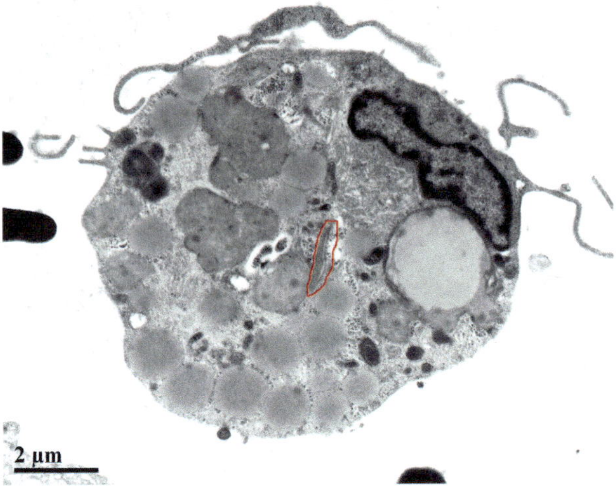

Fig. 3.98 Invasive lobular carcinoma cells of the breast, abundant secretory vesicles, bundles of tonofibrils (red circles) (TM, 12 000×)
Breast tumor cells contain many of the same secretory vesicles as normal breast cells, but the tumor cells have disturbed secretion and the composition of the secretion may be different. Under the electron microscope, these secretory vesicles are stained as grey, full vesicular structures, which are different from other phagocytic vesicles and secondary lysosomes. Because lysosomal vesicles are easily dissolved away by the solvents used for specimen processing, they often appear vacuolated, or they are themselves completely disintegrated and also show vacuolated changes. Breast tissue is rich in fibrous connective tissue, making some breast cancer cells also rich in tonofibrils

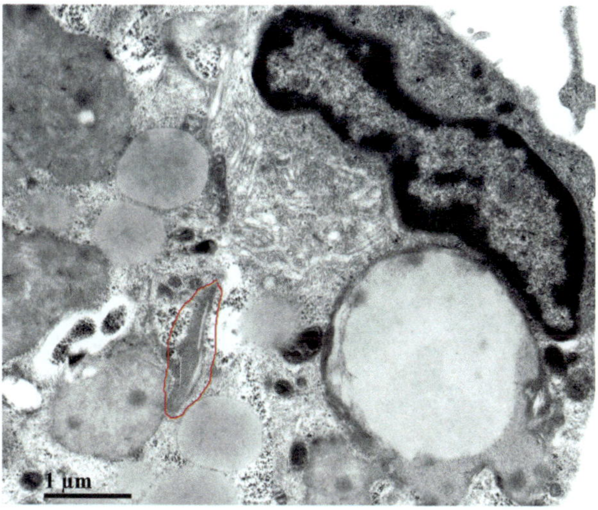

Fig. 3.99 Higher-magnification electron micrograph (close-up view of the area outlined in Fig. 3.98) showing bundles of tonofilaments (indicated by red circles) (TM, 25,000×)
Higher magnification reveals, in addition to secretory vesicles, abundant free ribosomes present both in clusters and individually, a well-developed endoplasmic reticulum system, small bundles of tonofibrils (indicated by red circles), and scattered mitochondria in the cytoplasm

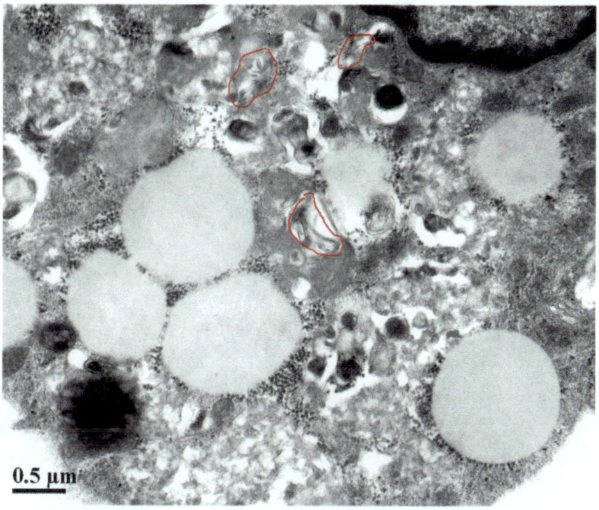

Fig. 3.100 Local magnified electron micrograph of a tumor cell (TM, 30000×)
In the red circle is the bundle of tonofibrils, which can be seen clearly after magnification of 30,000 times, and it is one of the skeletons that maintain the cell structure. It is clearly seen in squamous carcinoma, tumor of epithelial origin, and other epithelial tissues, and it can also be seen in other tissues or cells of other origins, with less quantity, but it can also be seen clearly in tumor cells occasionally, the tumor cells are adenocarcinomas, not squamous carcinomas, and the structure of clear bundles may be related to its origins, and the mammary gland is also rich in connective tissue responsible for the supporting role

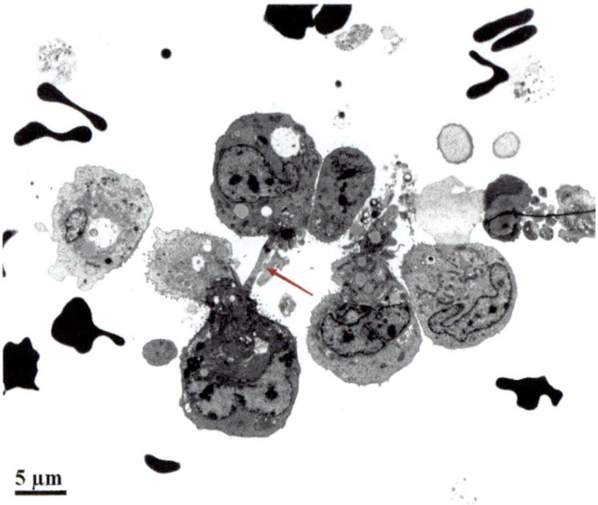

Fig. 3.101 Tumor cell intercellular junctions (TM, 3000×)

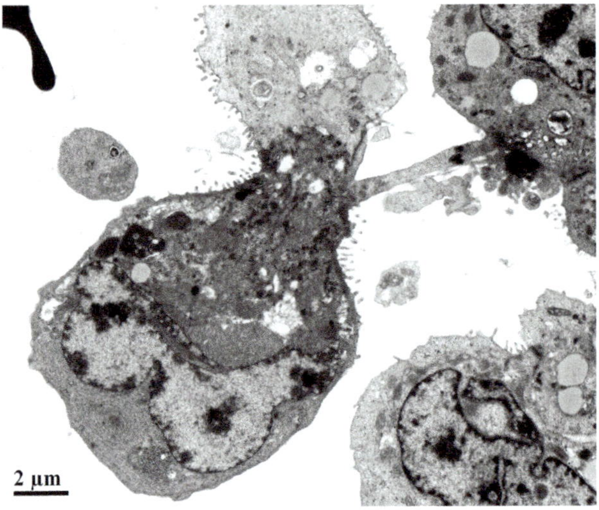

Fig. 3.102 Fig. 3.101 Electron micrograph of tumor cell intercellular junctions after local magnification (TM, 8000×)

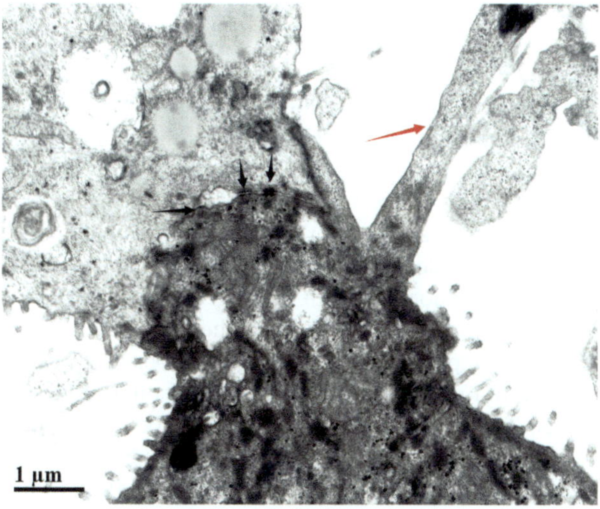

Fig. 3.103 Black arrows, from left to right, are tight junctions, intermediate junctions, and cellular desmosome, which together form the junctional complex; red arrows are luminal tubes, which are the passageways for intercellular transfer of materials (TM, 20,000×)

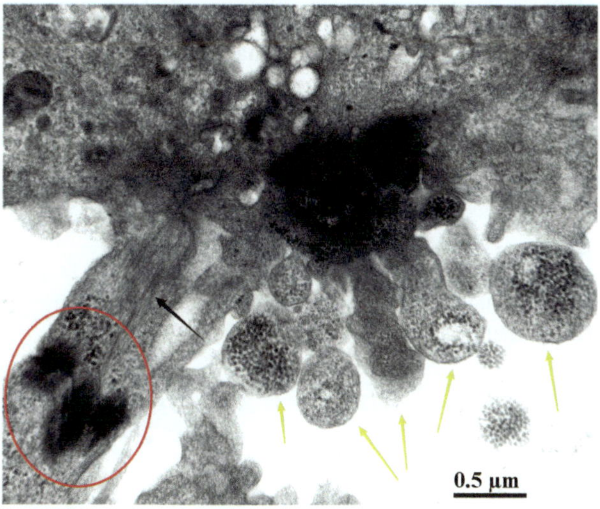

Fig. 3.104 Desmosome in the red circle and tonofibril indicated by the black arrow. Microvilli (yellow arrows) are clearly distributed in clusters, which are closely related to tumor metastasis and invasion, and there are a lot of free ribosomes in them, which are closely related to their synthesis of proteins and participation in the exchange of intercellular substances (TM, 40,000×)

Cellular junctions, usually common in studies of pathological tissues, are rarely seen intact in somatic cytomorphological studies. The cellular connections of the glandular epithelium have not only the core epithelial features of bridging granules, tonofibrils, and basal lamina but also luminal tubules, microvilli, junctional complexes, and secretory granules. Luminal tubules receive and store secretions for further externalization to ductal structures; microvilli have a role in digesting extracellular material, and the junctional complex promotes intercellular adhesion around the luminal tubules A. Exfoliated tumor cells also have these structures.

A quote from Brian Eyden's *The Ultrastructure of Human Tumors Applications in Diagnosis and Research.*

3.1.17 Adenocarcinoma of the Lungs

Case: Female, 74 years old, chronic disease. Two months ago, due to cough and sputum for 1 month to a hospital, chest CT showed irregular nodules in the upper lobe of the right lung; enlargement of the right lung hilar, right hilar, and mediastinum can be seen multiple lymph node enlargement; multiple inflammatory lesions in both lungs. There were pleural effusion and pericardial effusion on both sides. Bronchoscopy showed (1) inflammatory lesions in the lungs, (2) enlarged mediastinal lymph nodes, and (3) EBUS-TBNA. the pathology report of "EUBS-TBNA needle aspiration smear of group VII lymph nodes" showed a hemorrhagic background, with a few scattered free heterogeneous cells. IHC: CKpan (+), EMA (+), P63 (−), NapsinA (+), suggesting adenocarcinoma. Bronchoscopy was highly suspicious of malignant lesions, and the hospital suggested puncture to clarify the diagnosis. Chest CT examination showed nodules in the upper lobe of the right lung, suggesting tumor, multiple lymph node enlargement in the right hilar and mediastinal septum, multiple inflammatory lesions in both lungs, and pleural effusion and pericardial effusion on both sides. Cytopathological examination and diagnostic test: "Liquid-based smear of the left supraclavicular lymph node" found cancer cells. Smear of pleural fluid showed an increase in the number of nucleated cells, with a predominance of abnormal cells and a high cytoplasmic content, suggesting metastatic tumor cells. The cytomorphological examination of pleural fluid is shown in Figs. 3.105, 3.106, 3.107, 3.108, 3.109, 3.110, 3.111, 3.112, 3.113, and 3.114.

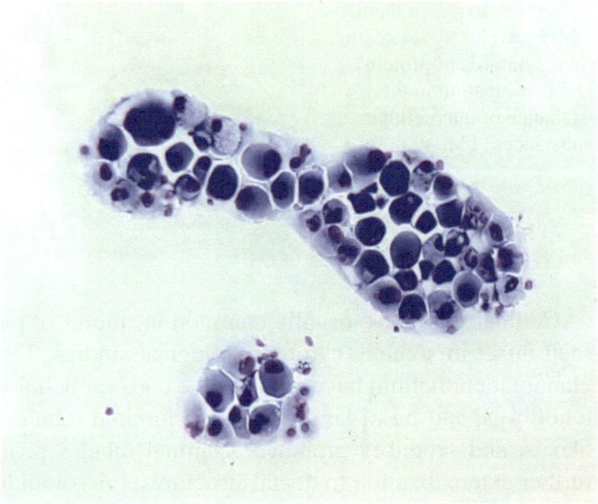

Fig. 3.105 Adenocarcinoma cells appear in clusters (Ritter-Giemsa stain, 400×)

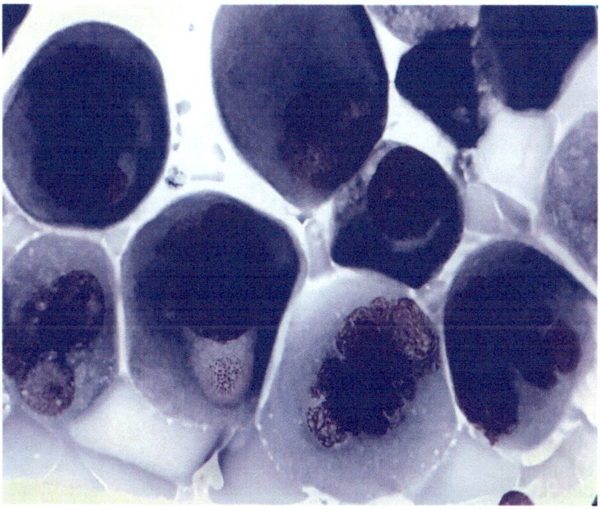

Fig. 3.106 Tumor cells with huge cell body, clear inner and outer plasma boundaries, malformed and deeply stained nuclei, autophagic vesicles in the cytoplasm, some of the tumor cells with obvious nuclear fragmentation, and massive proliferation of tumor cells, with a poor prognosis (Ritter-Giemsa staining, 1000×)

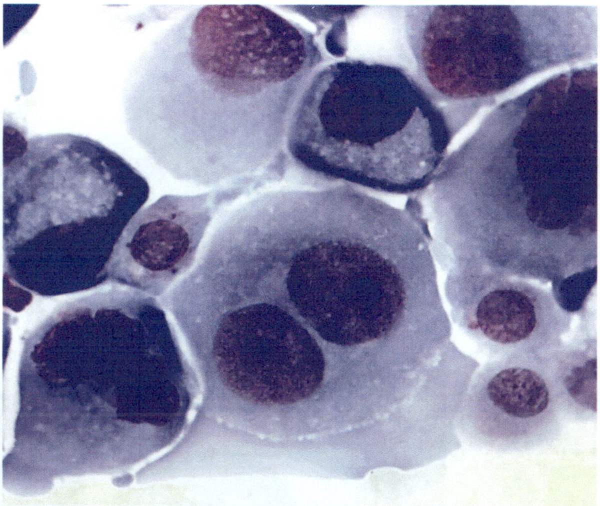

Fig. 3.107 Dinucleated tumor cells (Ritter-Giemsa stain, 1000×)
Nucleolus large and conspicuous, outer plasma resembling snail's carpal feet

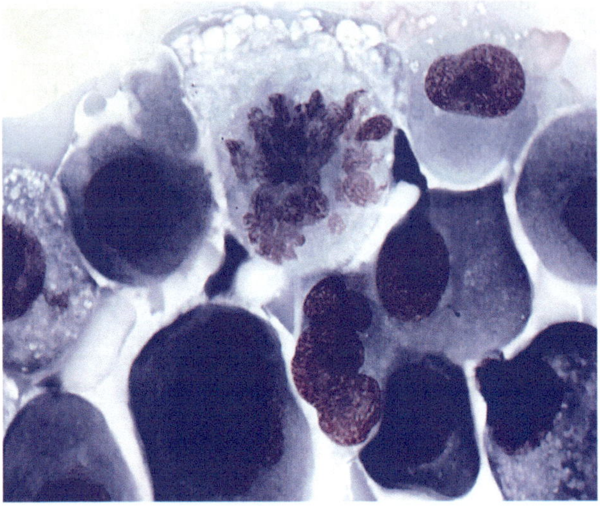

Fig. 3.108 Tumor cell mitotic figure with thick chromosomes (Ritter-Giemsa stain, 1000×)
The chromosomes in the mitotic figure are coarse, with inconspicuous polarity and disordered arrangement; nuclear fragmentation and malformation can be easily seen; vacuoles can be seen in the cytoplasm of some cells, suggesting that the cells are very active in autophagy or phagocytosis, which is in line with the characteristics of tumor cells

Fig. 3.109 Abnormal abundance of tumor outer plasma (Ritter-Giemsa stain, 1000×)The outer plasma of the tumor cells is contiguous, clear and translucent, without granularity

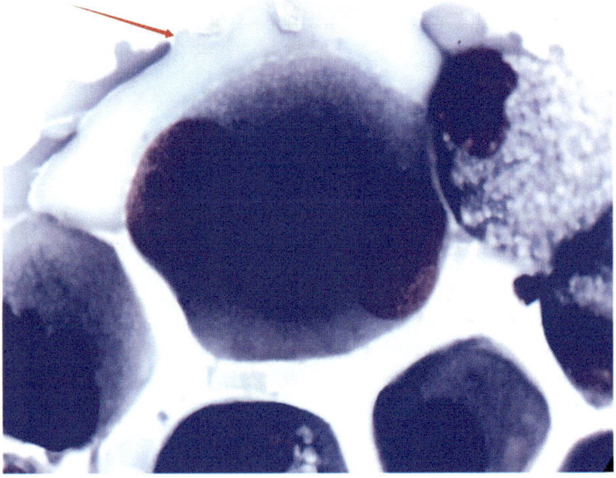

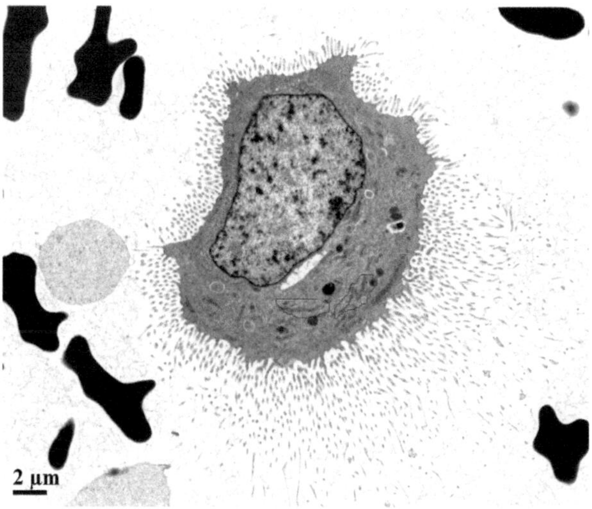

Fig. 3.110 Adenocarcinoma cell ultrastructure (TM, 5000×)
The cell body of the tumor cells is huge, the surrounding villi are long and obvious, arranged in a coronal pattern, and under the light microscope, the outer plasma is the outer plasma (without particles, like the carpals of a snail), and the villi are particularly developed, indicating that the tumor's adherence and colonization ability is very strong; the nuclear chromatin is loose, mainly euchromatin, indicating that the synthesis is powerful and active; a small number of tonofibrils can be seen in the cytoplasm (the red circle); and the glycogen is in a cluster (the blue circle). Free ribosomes are scattered and very abundant, giving the cell a grey-black color. The obvious increase of free ribosomes indicates that the protein synthesis of the tumor cells is powerful and different from that of normal cells, and the abundance of glycogen indicates that the organelles are very active and require a lot of energy. The presence of perinuclear fissures suggests that the cells have been in the pleural effusion for a long time and have a tendency to fragment and apoptose

Tumor spread can be divided into local infiltration, direct spread, lymph node metastasis, hematogenous metastasis and implantation metastasis, etc. Usually, early stage tumors are local infiltration and direct spread, while late-stage tumors often have distant metastasis through the blood or lymphatic tracts, and of course, it cannot be excluded that some individual malignant tumors can have distant metastasis in the early stage as well. A variety of factors play a role in the metastatic process, and extracellular microvilli or pseudopods play an important role in the process of adhesion and colonization. Under the electron microscope, different extracellular villi or pseudopods are different from each other, and they can also be used as one of the important features for identifying the cell type. The following are other types of cells that appear in the same specimen at the same time, which can be slightly distinguished by the extracellular pseudopods or villi, as shown in Figs. 3.113 and 3.114.

Usually, the detection rate is highest when the first chest and abdominal fluid is taken for examination, which contains more exfoliated cells, and more tumor cells can be seen, while it is difficult to find tumor cells in samples retained after multiple

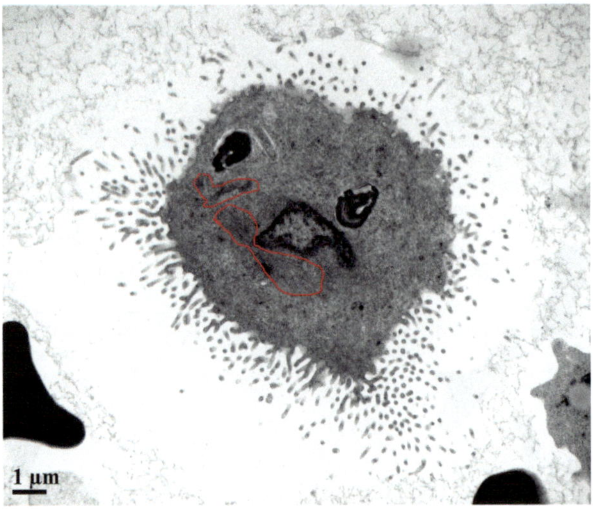

Fig. 3.111 Adenocarcinoma cells showing nuclear fragmentation (TM, 10000×)
The nucleus of this tumor cell was obviously fragmented, and the fragmented part of autophagy formed autophagic vesicles; tonofibrils (red circles) could be seen at the lower edge of the nucleus, and the extracellular microvilli were numerous and dense, and the cells were greyish-black due to free ribosomes and glycogen. Under the light microscope, only the inner and outer plasma can be seen, while under the electron microscope, the outer plasma can be seen as elongated and dense microvilli; the tonofibrils of the tumor cells cannot be identified under the light microscope but are clearly visible under the electron microscope; it is often not easy to distinguish the dense or loose chromatin under the light microscope, but under the electron microscope the chromatin is clear, and the nucleolus is obvious; free ribosomes and glycogen within the cytoplasm cannot be identified under the light microscope, while under the electron microscope, the abnormally increased number of free ribosomes and glycogen can make the cells appear greyish black. The number of mitochondria, Golgi apparatus, and other cellular contents and structural abnormalities are also clearly visible, which reflects the advantages of electron microscopy

Fig. 3.112 Presence of distinct autophagic vesicles in adenocarcinoma cells (TM, 5000×)
Adenocarcinoma cells with the presence of obvious autophagic vesicles, obvious nuclear aberrations, coarse nucleolus, abundant organelles in the cytoplasm, and thin and dense extracellular villi

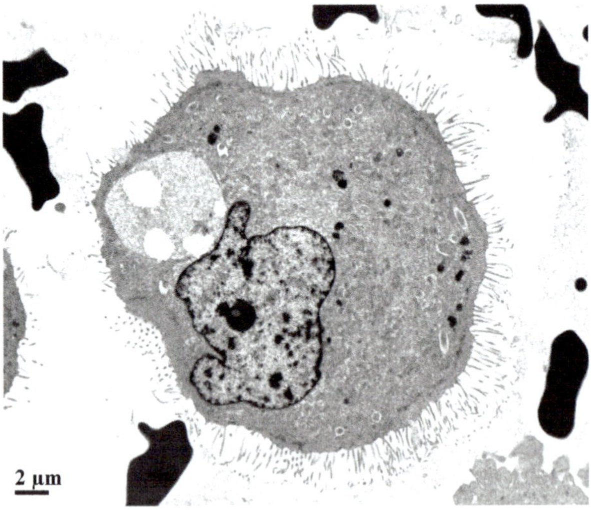

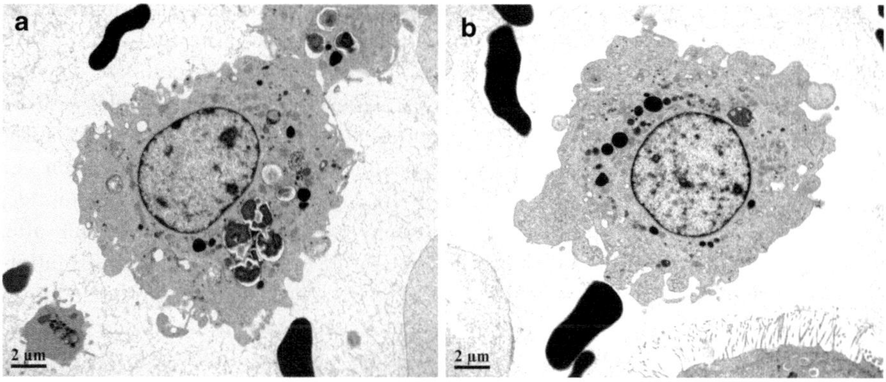

Fig. 3.113 Mesothelial cells (TM, 7000×)
The pseudopods of mesothelial cells are mostly warty or wrinkled, normal mesothelial cell nuclei are round, visible nucleolus, euchromatin is dominant, heterochromatin is distributed in a thin circle at the edge of the nuclear membrane. Mesothelial cells can also be seen in the phenomenon of pinocytosis, and the phagocytosis of macrophages is slightly different, the macrophage is more active phagocytosis after recognition of the target, mesothelial cells are more of a certain substance around the excess, and passive pinocytosis, so in the two mesothelial cells, can be seen in the pinocytosis of other cell residues and erythrocytes residues of black dense hemoglobin (hemosiderin particles)

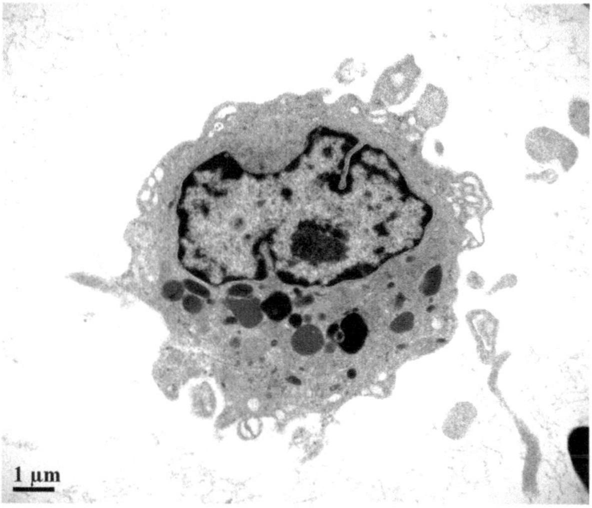

Fig. 3.114 Macrophages (TM, 12,000×)
Macrophages differ from mesothelial cells in that they are derived from monocytes, with irregular, folded, and twisted nuclei. The nuclei of activated macrophages are predominantly euchromatin, with nucleolus visible. In activated macrophages under light microscopy, Ritter staining was not easy to assess the degree of activation of the nucleus but could only roughly distinguish the degree of laxity and the presence or absence of nucleolus. Under the electron microscope, the pseudopods of activated macrophages were elongated and often curled into clusters, which were distinctly different from mesothelial cells; there were many secondary lysosomes in the cytoplasm, and a variety of particles containing catabolites could be seen; there were not many other organelles, and there were fewer free ribosomes, and the cells were lighter in color than those of all tumor cells except those of poorly differentiated carcinomas

punctures, especially those after washout treatment. It is necessary to communicate with the clinic that when sending samples for examination, the first examination items should be as comprehensive as possible, and when tumor cells are detected and further other examinations are needed, it is best to find other samples sent for examination in the same batch and add them as soon as possible. The same is true for electron microscopy experiments, the specimens of the same batch within 48h can be stored in refrigerator at 4~8°C for 7~14 days after fixation with glutaraldehyde, and it is better to do electron microscopy experiments within 1 week after fixation, and the cells of the stale specimen are prone to apoptosis or fragmentation, and the perinuclear gap is obvious. As the cells detached from the processed body fluids were mainly single cells, fewer clusters, compared with the tissue observed by electron microscopy, the intercellular bridges and junctional complexes were less visible, which was also a slight difference between the morphological judgment of histology and cytology.

Light and Electron Microscopy Comparison of Squamous Carcinoma

4

Dayong Gu and Ying Wang

4.1 Squamous Carcinoma

Squamous cell carcinoma, or squamous carcinoma for short, also known as epidermoid carcinoma, is a kind of malignant tumor originating from squamous epithelium, which is commonly found in areas covered by squamous epithelium, such as the skin, the oral cavity, the lips, the esophagus, the cervix of the uterus, the vagina, etc. Squamous carcinomas are not commonly seen on cytomorphological examination of body fluids. It commonly results from infiltration of adjacent sites or metastasis from distant tumors. Most commonly, squamous cell carcinoma of the esophagus or lung bronchus metastasizes in the chest or infiltrates the lung bronchus, and squamous cell carcinoma evolving from metaplasia of the bladder or renal pelvis can be seen in urine or ascites, and squamous cell carcinomas of the cervix or vagina in females tend to be found in ascites.

4.1.1 Lung Metastases from Malignant Tumors of the Esophagus (Squamous Carcinoma)

The patient is an 89 years old male with a chronic medical history. Eight months ago, he developed unexplained dysphagia and postprandial vomiting of nonprojectile, nonbloody gastric contents, without hematemesis, melena, chills, fever, abdominal pain, or diarrhea. The condition was not initially evaluated or treated. Subsequently, his dysphagia progressed to the point where he could only tolerate soft foods such as noodles, prompting him to seek medical attention five months ago. A non-contrast CT scan of the chest and esophagus revealed circumferential wall thickening of the lower esophagus, suggestive of a malignant lesion consistent with esophageal carcinoma,

D. Gu (✉) · Y. Wang
Laboratory Department, Shenzhen Second People's Hospital, Shenzhen, Guangdong, China

D. Gu, Y. Wang (eds.), *Morphological Diagnosis of Humoral Cells under Light and Electron Microscopy*, https://doi.org/10.1007/978-981-95-8668-4_4

and a follow-up gastroscopy was recommended. Gastroscopic findings indicated esophageal carcinoma (pending pathological confirmation), food residue, and chronic atrophic gastritis (C1). Pathological examination of an esophageal mucosal biopsy confirmed squamous cell carcinoma. After excluding contraindications, the patient underwent radiotherapy, receiving a total dose of 5040 cGy in 28 fractions over five weeks to the planning target volume (PTV) using 6 MV X-rays. He developed radiation esophagitis, which improved with symptomatic management, and grade II radiation-induced myelosuppression, which also resolved with treatment, leading to a period of stable disease. In February 2021, due to recurrent severe dysphagia, he successfully underwent a gastrostomy under local anesthesia. One month ago, he presented with chest tightness, significant shortness of breath, and hemoptysis, accompanied by a general decline in his condition, leading to hospital admission for further management. The cytomorphological examination of pleural fluid is shown in Figs. 4.1, 4.2, 4.3, 4.4, 4.5, 4.6, 4.7, 4.8, 4.9, and 4.10.

It has been reported in the literature that tumor cells emit some secretory vesicles outside the cell (these exosomes carry genetic information, substances that can lead to immune escape, or other substances), which are then engulfed by phagocytotic cells (e.g., macrophages or neutrophils). These macrophages or neutrophils then deliver these substances elsewhere, thus enabling the transmission of tumor information and metastasis. The transfer of genetic material from highly invasive tumor cells to less invasive tumor cells, or the abnormal growth and proliferation of normal cells through the action of cytokines or other signaling proteins, or the immune evasion of other tumor cells, so that the organism will no longer attack this type of tumor cells, and so on, can be achieved.

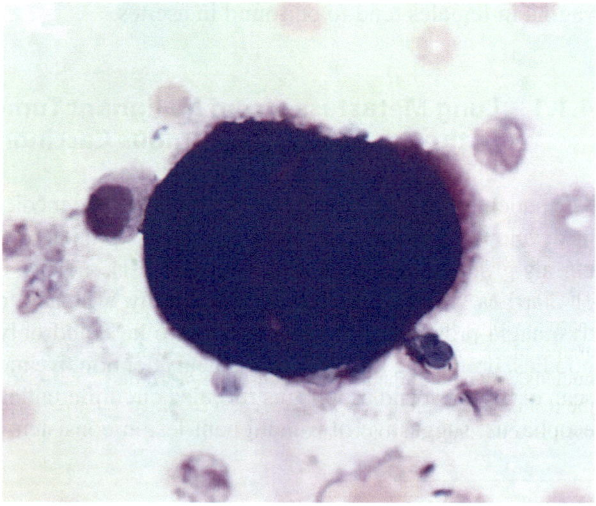

Fig. 4.1 Squamous carcinoma cells (Ritter-Giemsa Stain, 1000×) Cell bodies are small, with poorly defined intercellular boundaries, dark in color, and with conspicuous microvilli

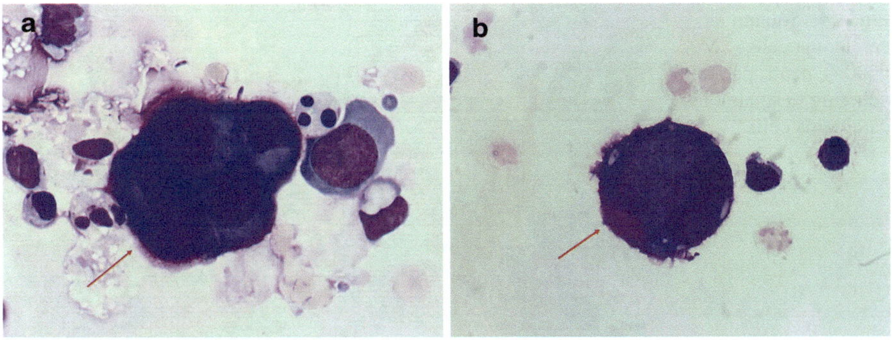

Fig. 4.2 Squamous carcinoma cells (Ritter-Giemsa Stain, 1000×)
Squamous carcinoma cells distributed in piles and clusters or scattered singly; microvilli short and dense, clustered or covered with extracellular membrane

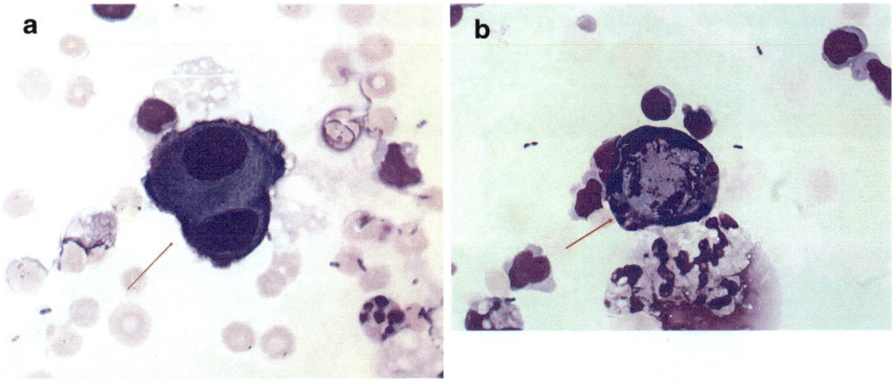

Fig. 4.3 Squamous carcinoma cells (Ritter-Giemsa Stain, 1000×)
Squamous carcinoma cells with pink villi, deeply stained nuclei, little cytoplasm, no visible secretory vesicles, single or clustered, and occasional apoptotic bodies. This patient has abnormal tumor markers

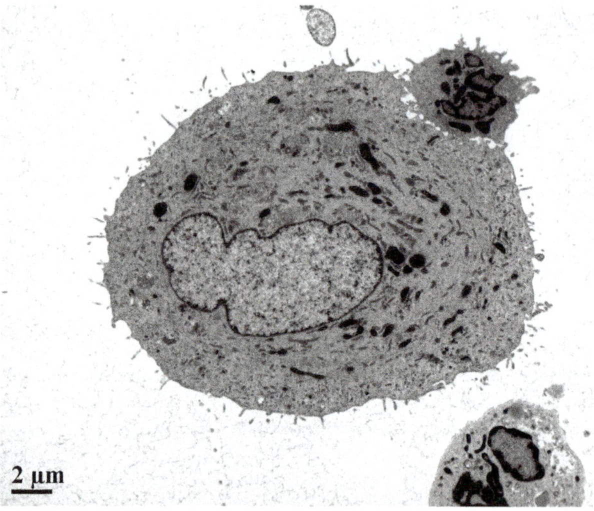

Fig. 4.4 Squamous carcinoma cells (TM, 6000×), neutrophils adhering to squamous carcinoma cells

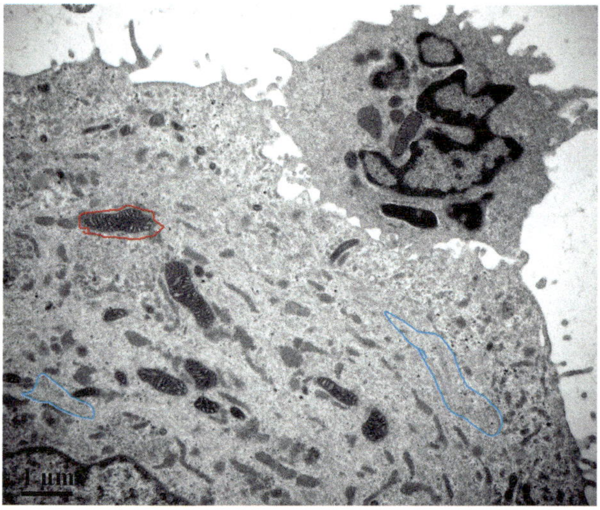

Fig. 4.5 A higher-magnification view of the area outlined in Fig. 4.4, showing ultrastructural details (TM, 15,000×)
Squamous carcinoma cells exhibit prominent mitochondria (indicated by red circles) with markedly thickened cristae, which may reflect organelle swelling due to suboptimal specimen preservation or, alternatively, particularly well-developed mitochondria with a proliferated inner membrane system in this cell type. Additionally, a distinct keratinized layer is visible at the cytoplasmic periphery (outlined in blue), accompanied by a significant increase in endoplasmic reticulum and free ribosomes within the cytoplasm. The extracellular microvilli are dense and prominent, in contrast to the shorter and less numerous pseudopods of neutrophils. Squamous carcinoma cells contain few secretory vesicles and only occasional phagocytic vesicles, which generally accounts for their smaller cell size compared to adenocarcinoma cells

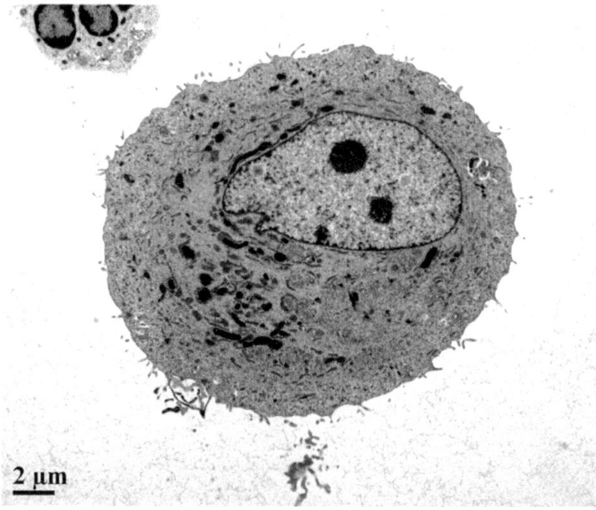

Fig. 4.6 Single, large squamous carcinoma cells (TM, 6000×) with distinct microvilli, pronounced perinuclear tonofibrils, and abundant mitochondria

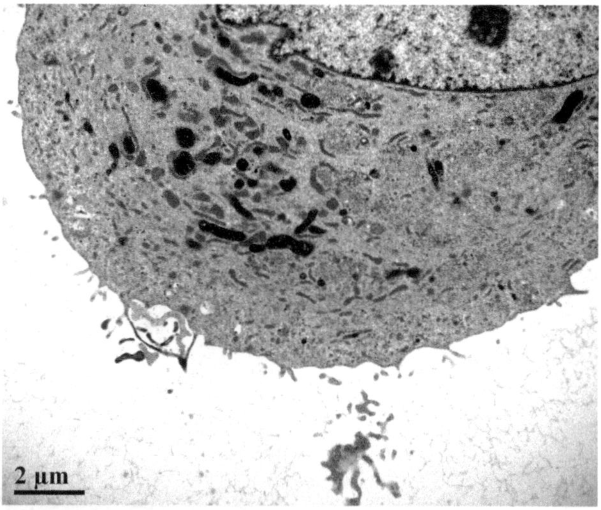

Fig. 4.7 A higher-magnification electron micrograph of the area outlined in Fig. 4.6 (TM, 10000×)
Mitochondria appeared electron dense, with clearly visible cristae, and were increased in number. The cytoplasm contained abundant free ribosomes and glycogen granules, and exhibited overall high electron density

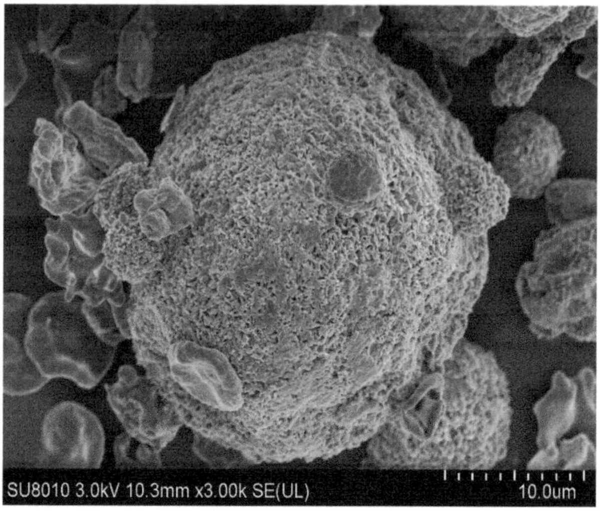

Fig. 4.8 Scanning electron microscopy of squamous carcinoma cells with abundant microvilli, three-dimensional, and strong attachment function

Fig. 4.9 Scanning electron microscopy of squamous cell carcinoma cells in piles and clusters, with cells tightly connected to each other by microvilli

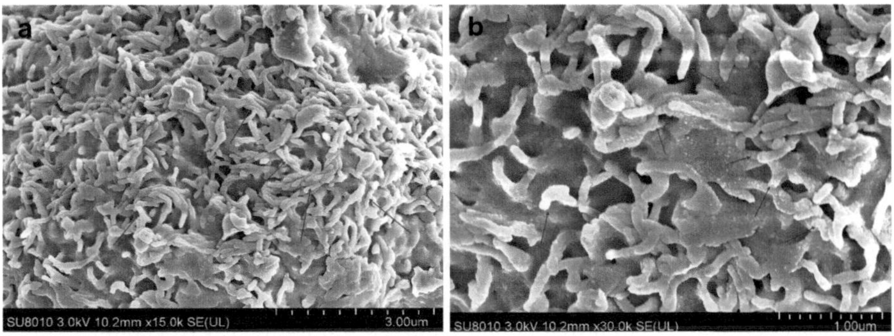

Fig. 4.10 Magnified squamous carcinoma microvillus
Unlike transmission electron microscopy, scanning electron microscopy can only observe the external structure of cells and their interrelationships. The villi of squamous cell carcinoma cells are very abundant at different magnifications, and they help the cancer cells to form clusters with each other, metastasize, adhere to and capture other cells, and transmit information, thus enabling them to thrive and proliferate in a malignant manner

4.1.2 Bladder Cancer (Squamous Carcinoma)

Case: Elderly patient, male, 82 years old, acute course. More than 10 years ago, physical examination found that the bladder was occupied, cystectomy was performed, and postoperative perfusion treatment was given (the specific process is unknown). Three months ago, the tumor recurred, and the pathology report of March 4, 2021, suggested that invasive carcinoma and the existing tissues were predominantly in the form of squamous cell carcinoma. On May 14, 2021, a follow-up CT showed: postoperative review of the bladder, local wall thickening, prostate calcification foci, retroperitoneal enlarged lymph nodes, and small lymph nodes on both sides of the iliac vessels and around the bladder. Thickening of the gastric sinus. Multiple cysts in both kidneys with ductal shadows. Some of the thoracolumbar vertebrae had small nodular high-density shadows. Re-excision was performed. Pathology report on May 17, 2021, suggested low differentiated invasive uroepithelial carcinoma; high-grade, extensive cancer embolism; invasion of intrinsic muscle layer; and large necrosis. After surgery, urinary retention occurred, and bilateral nephrostomy was performed. At present, the patient has no urinary pain, no urinary frequency, no urinary urgency, no abdominal pain, no abdominal distension, no nausea, no vomiting, no fever, and was admitted to the hospital for further treatment with the diagnosis of "malignant tumor of the bladder." He has been suffering from hypertension for several years and has been taking two timosartan tablets once a day, which he says is under good control. Urine routine results showed: urine protein ++ ↑, urine microalbumin +++ (>300) mg/L ↑, urine microalbumin/creatinine ++ (>30) mg/mmol ↑, urine occult blood +++ ↑, urine leukocytes +++ ↑, urine sediment erythrocyte count 6745.1/μl ↑, urine sediment leukocyte count 350.3/μl ↑, bacterial count 193.8/μl ↑, yeast positivity (≥1.0%), a positive (≥1.0%) count for a bladder tumor ↑, and yeast positivity (+) ↑. Coagulation results are as follows: prothrombin time 15.7s ↑, international normalized ratio 1.49 ↑, partial thromboplastin time 39.4s ↑, partial thromboplastin time ratio 1.43 ↑, D-dimer quantification 19,050.0 μg/L ↑, fibrinogen degradation 66,000.0 μg/L ↑. Biochemistry results showed the following: AST 67U/L ↑, AST 88U/L ↑, GTP 209U/L ↑, and alkaline phosphatase 335U/L ↑. Calcitoninogen test showed 1.47ng/ml ↑ and ferritin 1954.8μg/L ↑. Tumor markers (male) showed carcinoembryonic antigen 30.0 μg/L ↑, carbohydrate antigen 125 57.7 U/ml ↑, carbohydrate antigen 19-9 220.7 U/ml ↑, squamous cell carcinoma differentiation antigen 4.0 ng/ml ↑, cytokeratin 19 54.3 ng/ml ↑, and gastrin-releasing peptide precursor 116.2 pg/ml ↑. The cytomorphology of the ascites fluid is shown in Figs. 4.11, 4.12, 4.13, 4.14, 4.15, 4.16, 4.17, 4.18, and 4.19.

Cases of squamous cell carcinoma are less likely to appear than adenocarcinoma on cytomorphologic examination of body fluids. First, squamous cell carcinoma originates from epithelial tissue and is more common in areas covered by squamous epithelium (e.g., skin, etc.). Second, only well differentiated squamous cell carcinomas are better identified; poorly differentiated tumors of all kinds are difficult to recognize by morphology. Third, some tumors are adenosquamous carcinomas (a rare pathologic type of mixed carcinoma composed of squamous and adenocarcinoma. Microscopically, the tumor is composed of a mixture of adenocarcinoma and

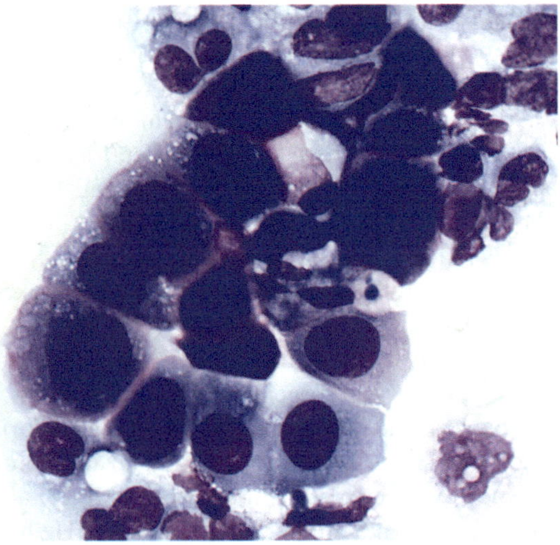

Fig. 4.11 Squamous cell carcinoma nests (Ritter-Giemsa stain, 1000×)
Typical squamous cell carcinoma cells are slightly smaller than adenocarcinoma cells, the nuclear deformity is not obvious, the chromatin is deeply colored, the edge of the cytoplasm has pink villi, and there may be a small number of small vacuoles, and the cell groups are often arranged in a queue-like or wall-like manner, but adenoid squamous cell carcinoma or adenosquamous carcinomas can also be seen, and it is not easy to distinguish atypical morphology from adenocarcinoma

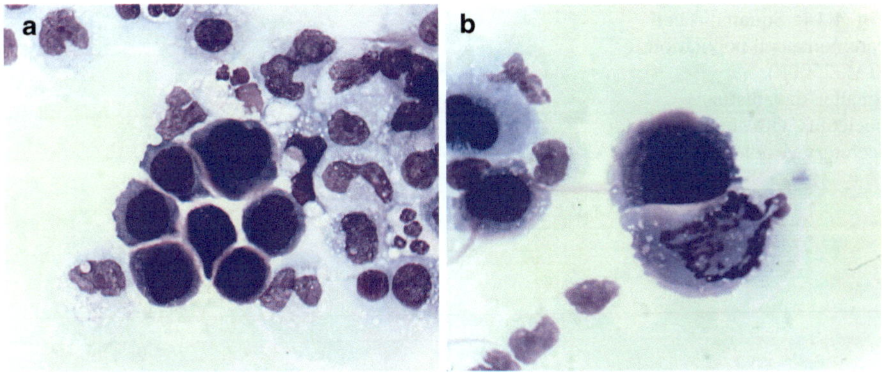

Fig. 4.12 Clustered squamous cell carcinoma cells (Ritter-Giemsa stain, 1000×)
Some squamous carcinoma cells show distinct inner and outer plasma, which is composed of dense microvilli under electron microscopy, while the distinction is not obvious under light microscopy

Fig. 4.13 Squamous cell carcinoma nests (Ritter-Giemsa Stain, 1000×)

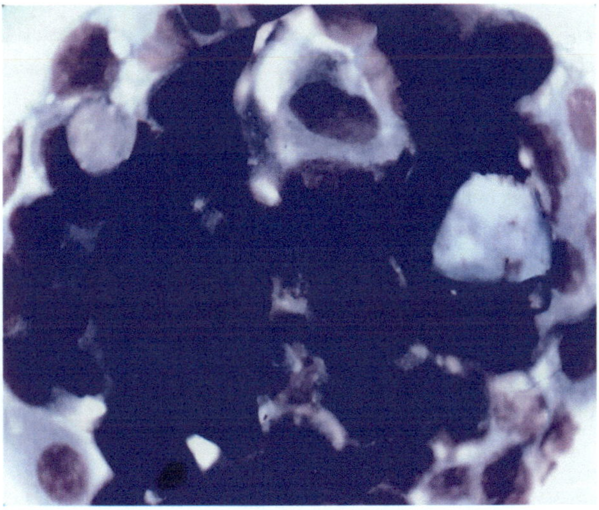

squamous carcinoma, with most of the two carcinoma components interspersed with each other or separated by fibrous mesenchyme and with each of the two components accounting for no less than 10% of the total amount of the cancerous tissue. It is common in males, and it is more likely to be found in older smokers and is located more often in the periphery of the lungs). The proportion of such cases is not low, and it is usually easy to recognize adenocarcinoma morphologically, while the squamous cell carcinoma cells mixed in are easy to be confused with mesothelial cells with heterogeneous nuclei, whose properties can only be determined by immunohistochemistry, so this part of squamous cell carcinoma is easy to be missed, and it can only be clarified by retrospective follow-up study combined with immunohistochemical pathology. Therefore, it is suggested that patients with various tumors should establish retrospective files to facilitate research follow-up.

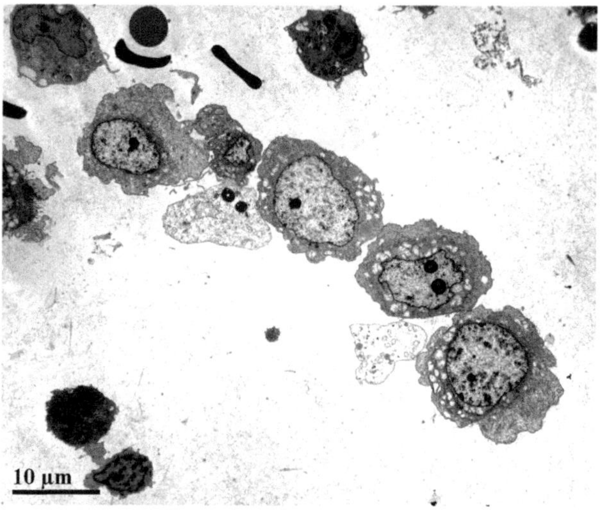

Fig. 4.14 Squamous cell carcinoma cell population (TM, 2500×)
Smaller size, distinct nucleolus with small secretory vesicles

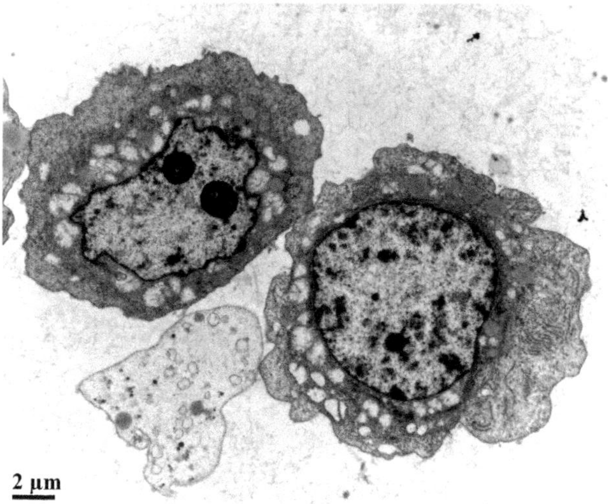

Fig. 4.15 Squamous cell carcinoma cells (TM, 6000×)
Tumor cells have loose chromatin, obvious whitening, swollen organelles in the cytoplasm, swollen endoplasmic reticulum, and obvious vacuoles, suggesting the beginning of apoptosis, which is commonly seen in excessive tumor cells or old specimens

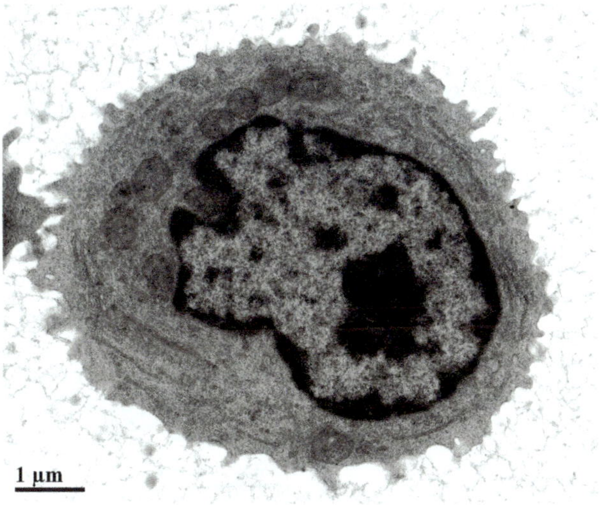

Fig. 4.16 Squamous cell carcinoma cells (TM , 20 000×)
The nucleolus is huge, euchromatin is dominant, mitochondria are abundant, swelling and degeneration are obvious, cytoplasm is rich in tonofibrils, cytoskeleton is clear, and microvilli are short and dense

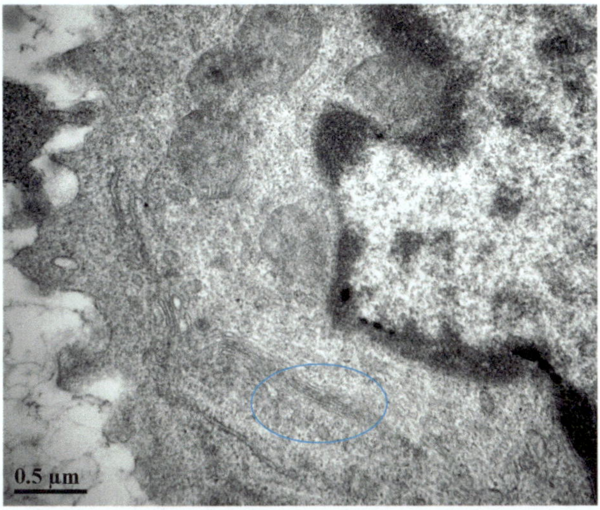

Fig. 4.17 A higher-magnification electron micrograph of the area outlined in Fig. 4.16 (TM, 40,000×)
Higher-magnification view showing prominent bundles of tonofibrils (indicated by blue circle) in well-differentiated squamous cell carcinoma

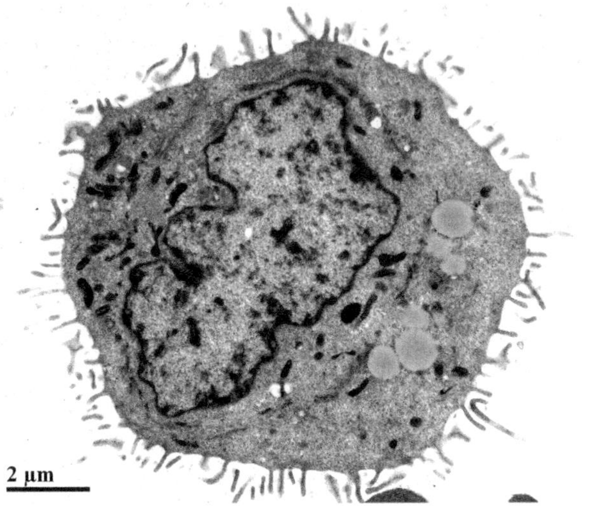

Fig. 4.18 Squamous cell carcinoma cells with follicles (TM, 12,000×) Nuclear aberrations evident, mitochondrial cristae thickened, deeply colored and numerous; secretory vesicles few and small; microvilli short, dense, and very well developed

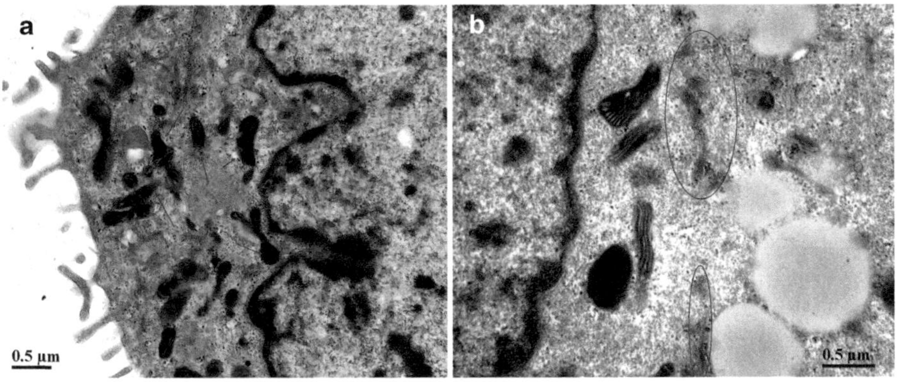

Fig. 4.19 Localized magnified electron micrographs (**a**. TM, 30,000×; **b**. TM, 40,000×) Typical squamous cell carcinoma cells. The mitochondria are well developed in the cytoplasm, and the cristae of the mitochondria are clearly visible after magnification of 30,000× and 40,000×, and the tonofibrils are in bundles (red circles). The microvilli on the cell membrane were short and dense, which were pink in Rhee-Ghee staining under the light microscope or light blue in the outer plasma. Structures such as junctional complexes and desmosomes are rare in individual squamous cell carcinoma cells but are commonly seen in histopathologic sections of well-differentiated tumors

Squamous cell carcinoma often occurs in areas covered by squamous epithelium, such as skin, lips, esophagus, larynx, cervix, vagina, and penis. In well-differentiated squamous cell carcinoma, stratified keratinized material called keratinized beads or cancer beads may appear in the center of squamous cell carcinoma nests, and a large number of intercellular bridges, junctional complexes, desmosome, and other structures can be seen in the inter-cellular area by electron microscopy, whereas poorly differentiated squamous cell carcinomas may have no keratinization, and the intercellular bridges and other structures may be few or absent.

Light and Electron Microscopy Comparison of Poorly Differentiated Tumor

5

Dayong Gu and Ying Wang

5.1 Poorly Differentiated Tumor

Poorly differentiated carcinoma is not a classification of tumors corresponding to adenocarcinoma, squamous cell carcinoma, or adeno-squamous carcinoma. Some textbooks classify these poorly differentiated and indistinguishable tumors as small cell carcinomas. According to the latest pathological classification of the World Health Organization (WHO) system, small cell carcinomas are one of neuroendocrine tumors, which corresponds to large cell carcinoma and so on. However, not all poorly differentiated and indistinguishable tumors are neuroendocrine tumors. Therefore, these morphologically indistinguishable low-differentiation tumors are classified as poorly differentiated carcinomas according to the degree of differentiation. The characteristics of poorly differentiated carcinomas include small cell body, large nucleo-cytoplasmic ratio, loose nuclear chromatin mainly dominated by euchromatin, large and prominent nucleolus, less cytoplasm, less various organelles, less free ribosomes, and light staining overall, which are not as strongly basophilic as other tumor cells. The cell attributes require pathological immunohistochemical analysis or genetic diagnosis. Morphologically, it only needs to indicate "suspected tumor cells" or "suspected poorly differentiated tumor cells." And further examinations are needed to clarify the cell attributes. Such tumor cells may be poorly differentiated adenocarcinoma, poorly differentiated squamous cell carcinoma, or other cancers.

D. Gu (✉) · Y. Wang
Laboratory Department, Shenzhen Second People's Hospital, Shenzhen, Guangdong, China

D. Gu, Y. Wang (eds.), *Morphological Diagnosis of Humoral Cells under Light and Electron Microscopy*, https://doi.org/10.1007/978-981-95-8668-4_5

5.1.1 Poorly Differentiated Carcinoma of the Pleura, Possibly Malignant Mesothelioma

Case: Elderly patient, male, 81 years old, chronic course. On May 17, 2021, the chest CT report showed a subpleural soft tissue mass in the mediastinum of the upper lobe of the right lung, which was considered to be a malignant tumor involving the adjacent mediastinal pleura, mediastinum, and the posterior wall of the trachea, and enhancement and fiberscope examination were suggested. After admission, cytopathological examination and diagnosis were carried out, and no cancer cells were found in the liquid-based smear of pleural effusion. 2021 PET report on May 24 suggested that the right pleura was thickened with multiple cords and nodules, and the metabolism of FDG was increased, and it was considered that malignant lesions and malignant mesothelioma might be present. He was given closed chest drainage, fluid replacement, anti-infection, and cough and sputum treatment and was discharged after symptoms improved. He came to the outpatient clinic for review and was admitted to the hospital for further treatment of "malignant pleural effusion." The patient's condition progressed rapidly until his death, and the reports of several pathological liquid-based cytology examinations were negative, but the morphological examination of the body fluid cells in the pleural effusion revealed tumor cells, which were few in number but had a typical morphology. The morphologic findings of the body fluid are shown in Figs. 5.1, 5.2, 5.3, 5.4, 5.5, 5.6, 5.7, and 5.8.

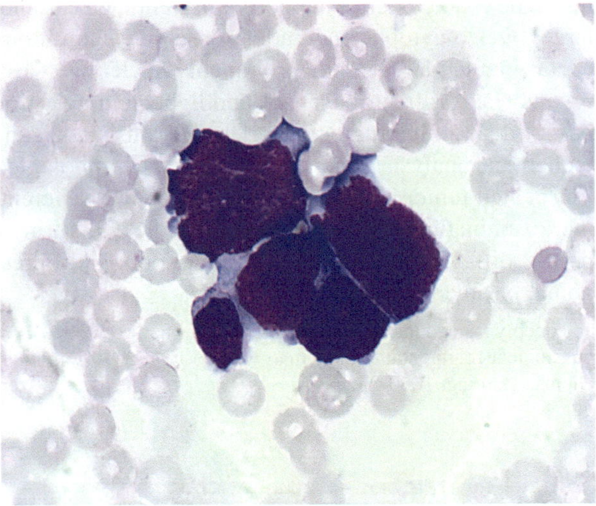

Fig. 5.1 Poorly differentiated pleural cancer cells (*Wright-Giemsa staining, 1000×*) Medium cell body, large nucleo-cytoplasmic ratio, almost naked nucleus-like, visible nuclear petal-like deformity, little cytoplasm, light blue, cells adhering in clusters, unclear cytoplasmic boundaries

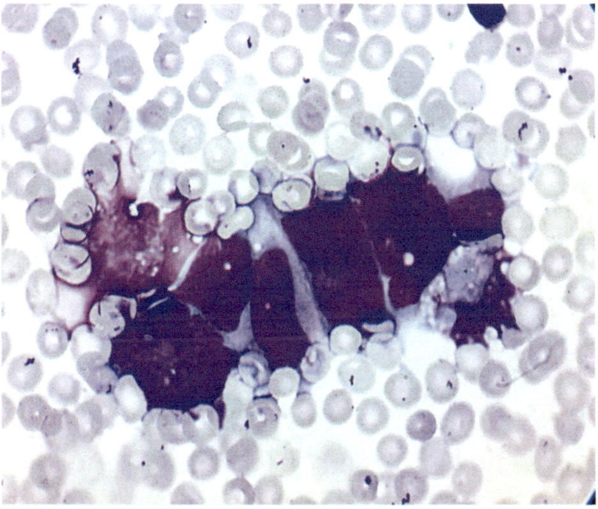

Fig. 5.2 Cohort-like arrangement of poorly differentiated tumor cells (Ritter-Giemsa stain, 1000×)
Individual poorly differentiated tumor cells should be differentiated from the blast cells of hematologic tumors. Usually, blast cells from hematologic tumors rarely adhere in clusters, whereas poorly differentiated cancer cells often appear in clusters. Individual poorly differentiated tumors that are slightly larger and morphologically indistinguishable suggest further flow cytology, immunohistochemistry, genetics, or molecular biology to differentiate them

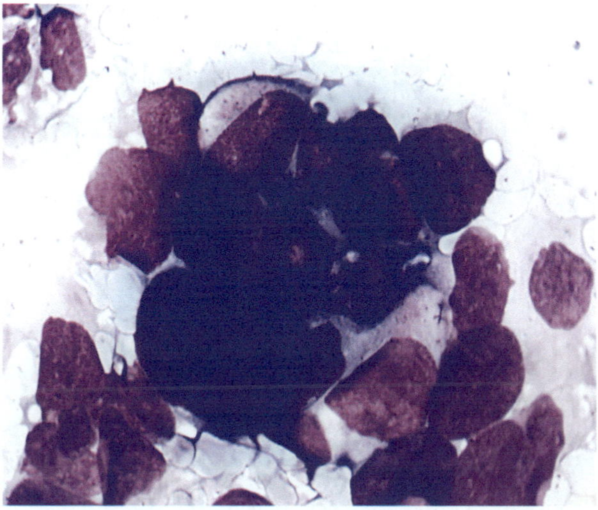

Fig. 5.3 Clusters of poorly differentiated tumor cells (Ritter-Giemsa stain, 1000×)
Naked nucleus-like tumor cells are often seen; note the differentiation from smudge cells, if atypical, typical tumor cells need to be found before making the diagnosis. Poorly differentiated tumors mostly see naked nucleus-like tumor cells, suggesting rapid differentiation and proliferation and poor prognosis

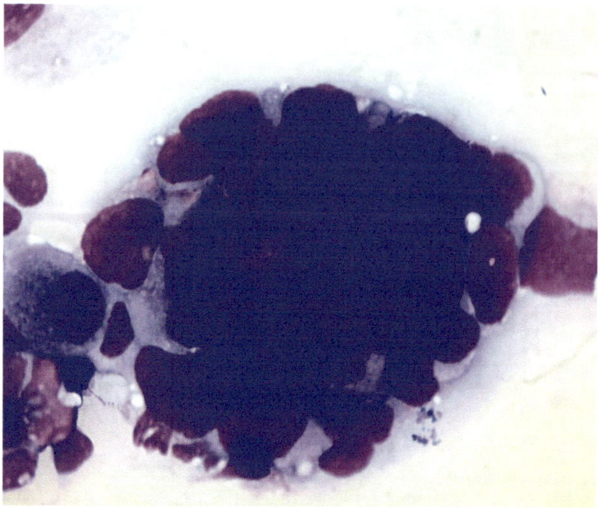

Fig. 5.4 Clusters of poorly differentiated tumor cells (Ritter-Giemsa stain, 1000×)
At the end of the smear, on the "shoreline," there are obvious clusters of tumor cells, with disordered arrangement, with unclear plasma boundaries, almost only the nucleus is visible, the amount of cytoplasm of the cells is small, the nuclear chromatin is loose, and occasional nucleoli are seen

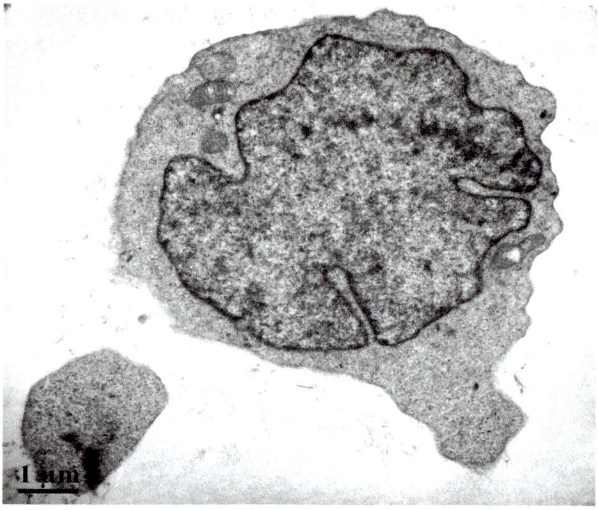

Fig. 5.5 Poorly differentiated tumor cells (TM , 15 000×)
There are very few organelles, only a few mitochondria in the cytoplasm; the nucleus is huge, malformed, with sparse chromatin, dominated by euchromatin; there are few free ribosomes in the cytoplasm, little glycogen, and the cell coloration is light. Lymphocytes also have few organelles and light cytoplasmic staining, and it is important to note the difference between the two: lymphocytes have aggregated nuclear chromatin, predominantly heterochromatin, and usually have smaller cell body than less differentiated tumor cells

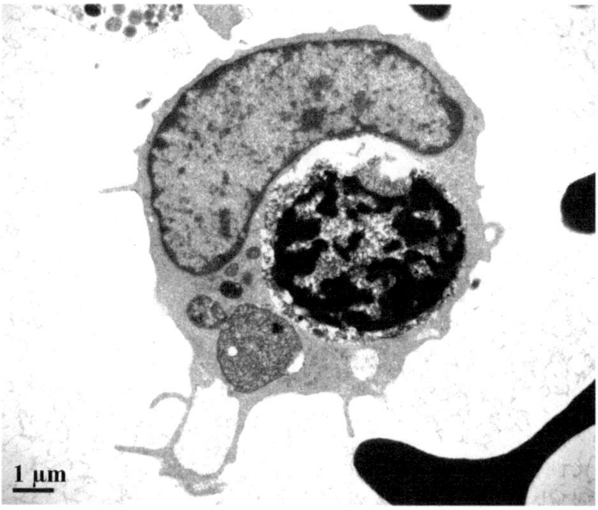

Fig. 5.6 Phagosomes of poorly differentiated tumor cells (TM , 12 000×)
Cells phagocytosed by tumor cells are apoptotic, chromatin is clumped, cell membranes are fragmented and blurred, and other phagocytic vesicles are seen in the cytoplasm of tumor cells

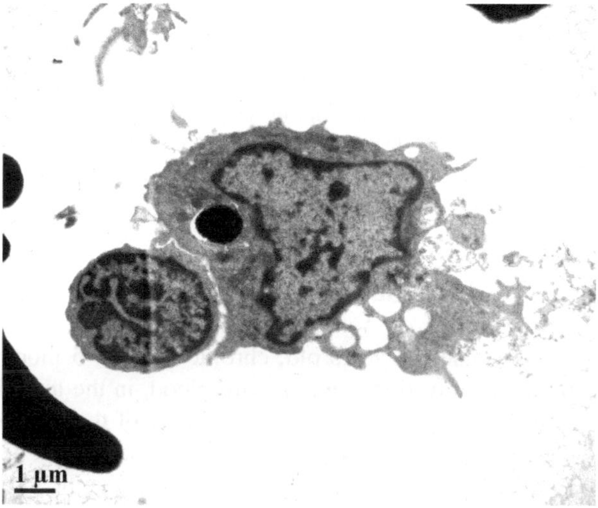

Fig. 5.7 Poorly differentiated tumor cells and lymphocytes adhering together (TM, 12 000×)
Tumor cells interact with lymphocytes and may be in the process of phagocytosis. Tumor cells have incomplete cell membranes and are close to apoptosis. Tumor cells proliferate too rapidly, and apoptosis is common, especially in body fluids, where prolonged immersion and other factors act to shatter cells commonly

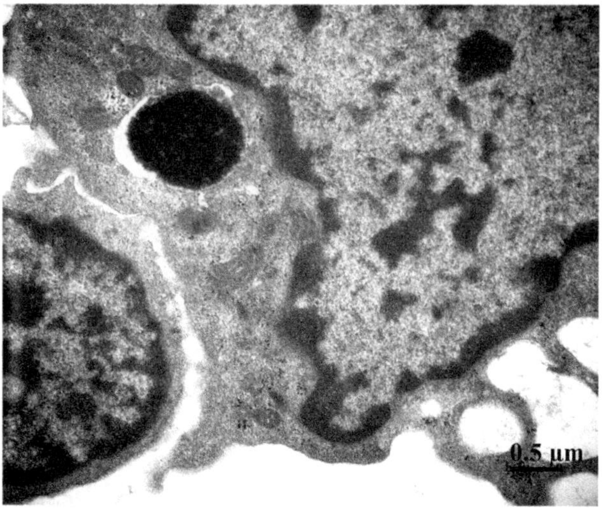

Fig. 5.8 Local magnification of poorly differentiated tumor cells (TM, 30 000×)
The cell has a low cytoplasmic volume, with a few scattered mitochondria and a phagocytic vesicle visible. Compared with other tumor cells, poorly differentiated tumor cells are smaller in size and need to be distinguished from normal lymphocytes and blast cells in hematologic tumors. In electron microscopy mature lymphocytes are slightly smaller than poorly differentiated tumor cells, nuclear chromatin is predominantly heterochromatin, nucleoli are rare, the cell body is small, and there are few organelles and a few mitochondria and occasional phagocytic vesicles. Activated lymphocytes, lymphoma cells and hematological blast cells are morphologically similar to poorly differentiated tumor cells, which is relatively difficult to identify, and the arrangement of cells can be different: poorly differentiated carcinoma often appears in clusters of several cells, with unclear plasma boundaries between the cells; other cells are often dispersed, with rare aggregation of lymphocytes, and even if clusters of lymphocytes are formed, the gaps between the cells are clearly visible and are mostly intact and independent. Flow cytology, immunohistochemistry, and molecular biology are needed to differentiate lymphoma or lymphatic system related diseases. Therefore, it is recommended to do other confirmatory experiments in morphologically suspicious cases, so as to avoid morphological misjudgment

5.1.2 Small Cell Carcinoma of the Lungs

Case: Elderly patient, male, 72 years old, chronic disease. 5 months ago, due to cough, cough sputum aggravation, sputum with blood, in the local hospital, chest enhancement CT report suggests: the right upper lobe of the lung tumor with the right lung in the upper lobe of the atelectasis, mediastinal lymph node metastasis. He was given anti-inflammatory drug treatment, the specific drugs used were unknown, and the effect was not good. On January 7, 2021, bronchoscopy was performed, and the pathology report of the brush smear of the middle section of right bronchus suggested that cancer cells were found, and small cell carcinoma was considered, and radiotherapy was recommended in the consultation, which was rejected by the patient and his family at that time. On April 20, 2021, after excluding contraindications, radical radiotherapy was performed. After the radiotherapy was finished, there was no obvious adverse reaction. One month ago, the patient had chest tightness, shortness of breath, cyanosis of lips and mouth, a little pain in chest and back, no obvious abdominal pain, no chest tightness, shortness of breath, fear of cold, fever, and other discomforts, so she was admitted to the hospital for further treatment of "malignant tumor of the lung." After reviewing the chest CT, the right pleural effusion was detected, and on May 20, 2021, ultrasound-guided puncture and drainage was performed. On May 24, 2021, cytopathological examination was performed to diagnose the disease, and cancer cells were detected in the liquid-based smear of pleural effusion. After exclusion of contraindications, cisplatin 60 mg thoracic perfusion chemotherapy was administered on May 25, 2021, with no obvious adverse reactions to radiotherapy. The cytologic examination of pleural fluid is shown in Figs. 5.9, 5.10, 5.11, 5.12, 5.13, and 5.14.

Small cell carcinoma belongs to neuroendocrine neoplasm, and secretory vesicles can be seen in tumor cells; some tumor cells have obvious nuclear deformity, and mitotic figure can also be seen in tumor cells, and if the number of mitotic figure increases, it suggests that the growth and differentiation of the tumor cells are active, and the prognosis is not good. The mitotic figure of tumor cells is usually larger than that of normal cells, and the number of chromosomes is large, some of which belong to euploid, and some belong to aneuploid, which can be seen that the development and division are not balanced, which is related to the malignant proliferation of the tumor cells and not subject to normal regulation. According to the description of proliferation index of neuroendocrine neoplasm by the WHO, the level of proliferation index is closely related to the prognosis of the tumor. The proliferation index is a criterion based on the classification of mitotic figures and Ki-67 index into three grades: G1 is <2 mitotic figures/10HPF and/or Ki-67 index ≤2%; G2 is 2 ~ 20

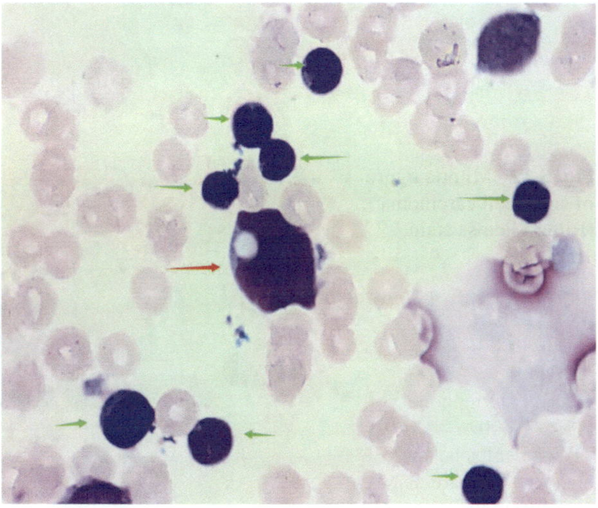

Fig. 5.9 Green arrows are mature lymphocytes and red arrows are single small cell carcinoma cells. Tumor cells are of medium size, with loose chromatin and low cytoplasmic volume (Ritter-Giemsa stain, 1000×)

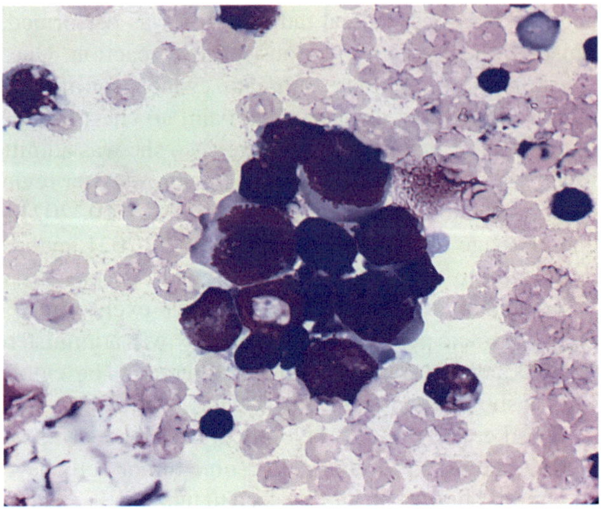

Fig. 5.10 Clustered small cell carcinoma cells (Ritter-Giemsa stain, 1000×) with obvious nuclear malformations, large nucleo-cytoplasmic ratio, and naked nucleus-like tumor cells

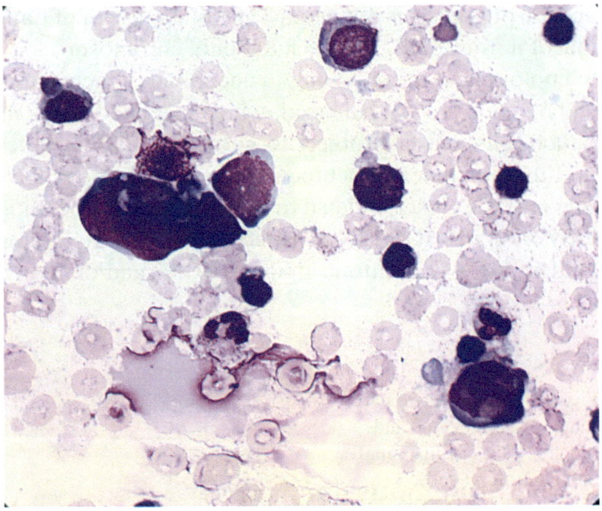

Fig. 5.11 Small cell carcinoma cells (Ritter-Giemsa stain, 1000×) Small cell carcinoma cells are clustered or scattered singly. Slightly larger in size than neutrophils and lymphocytes, with obvious nuclear malformations

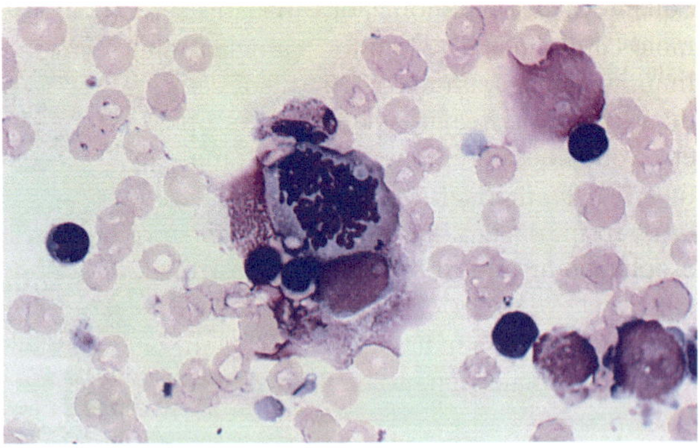

Fig. 5.12 Mitotic figure of small cell carcinoma (Ritter-Giemsa stain, 1000×)

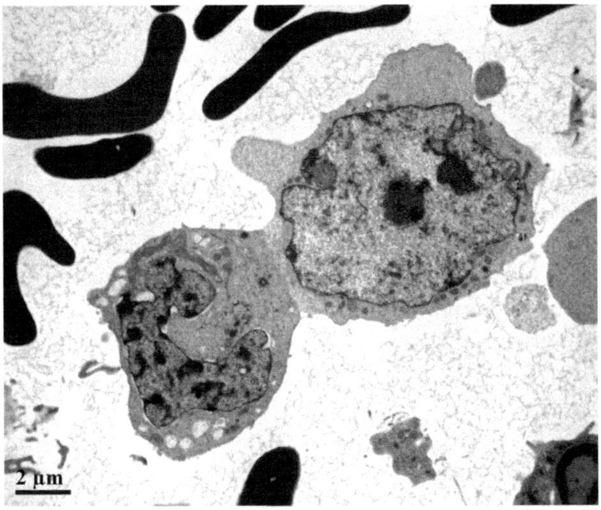

Fig. 5.13 Small cell carcinoma cells (TM, 8000×)
Two tumor cells are tightly contiguous, with malformed nuclei, large nucleoli, and predominantly euchromatin; low cytoplasmic volume and few organelles; light coloration

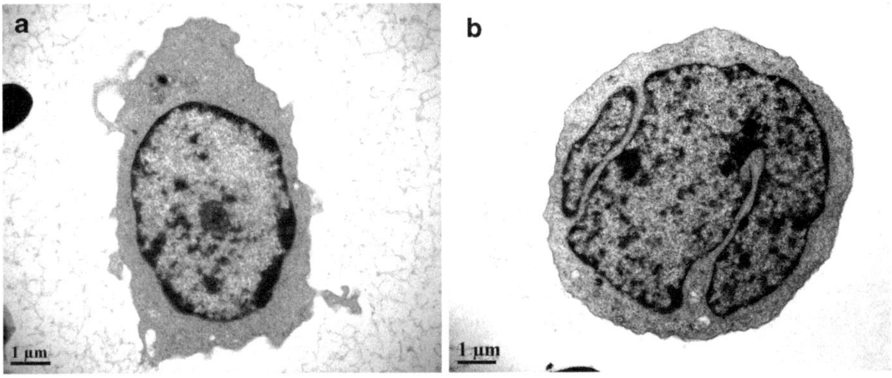

Fig. 5.14 Poorly differentiated tumor cells (TM, 15,000×) with large, loose nuclei
Poorly differentiated tumor cells have large and lax nuclei, pronounced nucleoli, a small amount of cytoplasm, and few organelles, making it difficult to differentiate the source, and definitively identifying the source and category requires immunohistochemistry, genetics, and molecular biology. Morphologic initial screening, suggestive of clinical abnormalities, requires further examination to clarify the diagnosis. Morphologic diagnosis is rapid, sensitive, inexpensive, and plays an important role in early diagnosis and screening. It is strongly recommended that body fluid be sent for cytomorphologic examination whenever fluid is present or when the cellular composition of the body fluid needs to be known. Body fluid cytomorphology can check not only the classification of cells but also bacteria, fungi, parasites, and crystals

mitotic figures/10HPF and/or Ki-67 index 3% ~ 20%; and G3 is ≥20 mitotic figures/10HPF and/or Ki-67 index >20%. This classification requires counting at least 50 high magnification fields of view (1 high magnification field of view = 2 mm^2); MIB antibody is used, and a minimum of 500–2000 cells are counted positive for the areas of strongest nuclear marker staining (hot spots). If there is a discrepancy between the mitotic figure grading and the Ki-67 index grading, the higher grade is graded. There is evidence that this grading scheme applies to neuroendocrine neoplasm of the stomach, duodenum, and pancreas, with the exception of the rest of the intestine. As for other organs and tissues, the applicability of this grading scheme can be determined by referring to the WHO pathologic criteria for other sites. In conclusion, a marked increase in mitotic figures is indicative of rapid proliferation and a poor prognosis.

5.1.3 Gastric Poorly Differentiated Adenocarcinoma with Brain Metastasis

Patient, female, 58 years old, with a chronic disease course. In March 2017, she underwent radical gastrectomy for gastric cancer (total gastrectomy with Roux-en-Y anastomosis, D2 lymph node dissection). Pathology revealed a diffusely infiltrating, poorly differentiated adenocarcinoma of the gastric antrum with focal signetring cell carcinoma (Lauren classification: diffuse type) and lymph node metastasis. Postoperatively, she received 6 cycles of chemotherapy with paclitaxel plus tegafur, gimeracil, and oteracil potassium (S-1). In July 2017, she underwent high-intensity focused ultrasound (HIFU) ablation for a metastatic lesion in segment IV of the liver. A permanent ileostomy was performed on December 3, 2019. During the surgery, a large amount of pale red ascites was noted in the abdominal cavity; the visualized small intestine, colonic mesentery, and serosal layer appeared grossly unremarkable, while the pelvic floor and rectum were rigid, though their surfaces were smooth. Cytopathological examination on December 5, 2019, of ascitic fluid via liquid-based cytology confirmed the presence of malignant cells. A follow-up evaluation on September 16, 2020, indicated disease progression. The patient declined further chemotherapy. Based on genetic testing which identified a PDGFR mutation, targeted therapy with dasatinib was initiated. After two weeks, the patient's symptoms showed no improvement, and tumor markers had increased, suggesting uncontrolled disease; a maintenance treatment strategy was therefore planned. In the week prior to presentation, she developed nausea and vomiting, preventing oral intake; the vomitus consisted of white mucus, occurring 10–20 times per day, raising high suspicion for leptomeningeal metastasis and indicating a poor prognosis. On June 4, 2021, she underwent gamma knife radiosurgery.

Subsequently, she experienced persistent nausea, vomiting, dizziness, and epileptic seizures. A lumbar puncture performed on June 7, 2021, confirmed leptomeningeal metastasis. She is currently admitted for further treatment under the diagnosis of gastric cancer.Usually,the cellularity in CSF is typically lower than in other body fluids. Using a cytocentrifuge for slide preparation enhances tumor cell detection. In this patient, routine CSF analysis showed a nucleated cell count of only 15/μL, and the cytomorphology of the fluid was examined after concentration by cytocentrifugal smear machine, as shown in Figs. 5.15, 5.16, 5.17, 5.18, 5.19, 5.20, 5.21, 5.22, and 5.23.

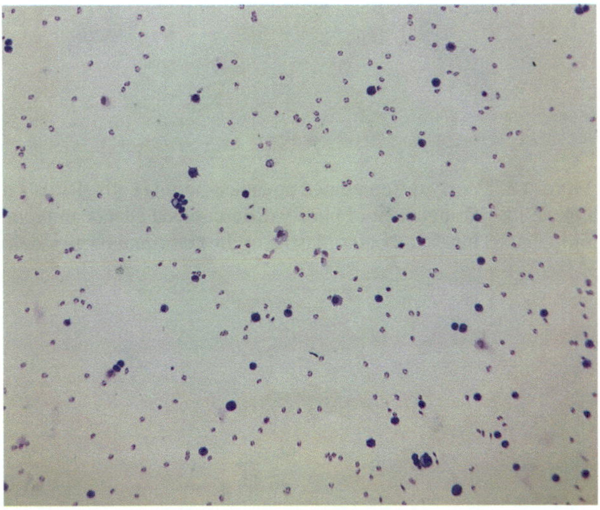

Fig. 5.15 Cerebrospinal fluid (Ritter-Giemsa Stain, 100×)
Tumor cells can be seen scattered distribution, cell clusters are rare, and the size of the neutrophils is 1.5–2 times

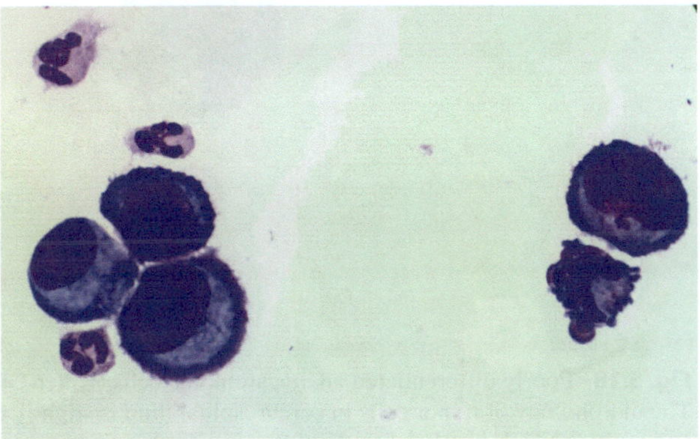

Fig. 5.16 Poorly differentiated adenocarcinoma cells in cerebrospinal fluid (Ritter-Giemsa stain, 1000×)
Poorly differentiated adenocarcinoma cells in cerebrospinal fluid with large cell bodies, dark staining, and distinct microvilli

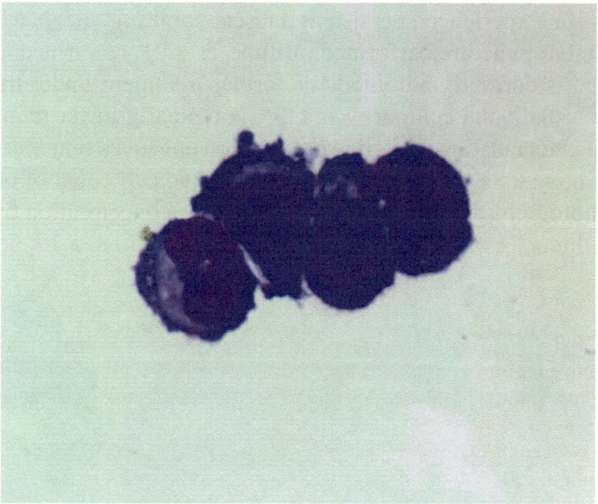

Fig. 5.17 Poorly differentiated adenocarcinoma cells (Ritter-Giemsa stain, 1000×)
Queue-like arrangement of poorly differentiated adenocarcinoma cells with conspicuous nucleoli, pronounced inner and outer plasma of the cells, and deeply stained cell bodies

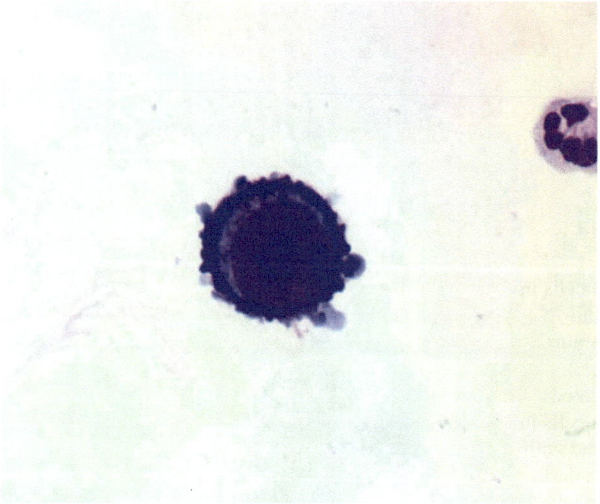

Fig. 5.18 Poorly differentiated adenocarcinoma cells (Ritter-Giemsa stain, 1000×)
The morphology of tumor cells in cerebrospinal fluid is slightly different from that in pleural fluid and ascites, which is related to the difference in their internal environments. The colloid osmotic pressure in the cerebrospinal fluid is low, while the crystal osmotic pressure is high, which increases the brittleness of cell membranes, and the cells are easy to break up and are more dispersed and seldom aggregated into a group, so that the cells are more spread out, more regular, and the structure is clearer

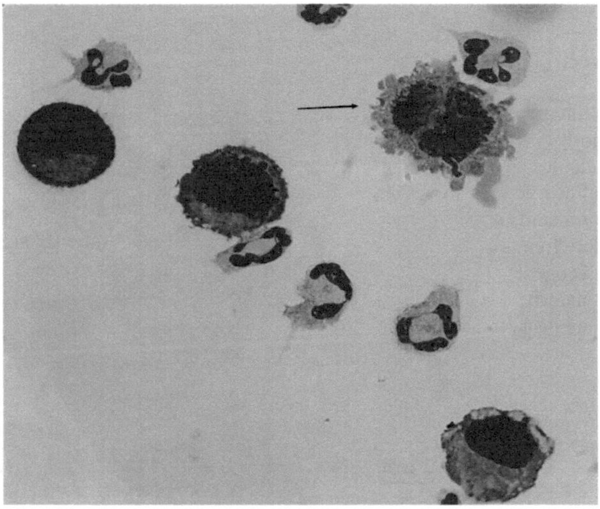

Fig. 5.19 Abnormal mitotic figure of poorly differentiated adenocarcinoma cells (Ritter-Giemsa stain, 1000×)
The cell bodies of the tumor cells show numerous pink microvilli, warty projections, and unilateral clusters of pink microvilli. The cells are darkly stained throughout, with loose nuclear chromatin, one to several large nucleoli, and pink eosinophilic areas on one side of the cytoplasm. The cells referred to by the black arrow show asynchronous development of mitotic figures, which are unevenly distributed

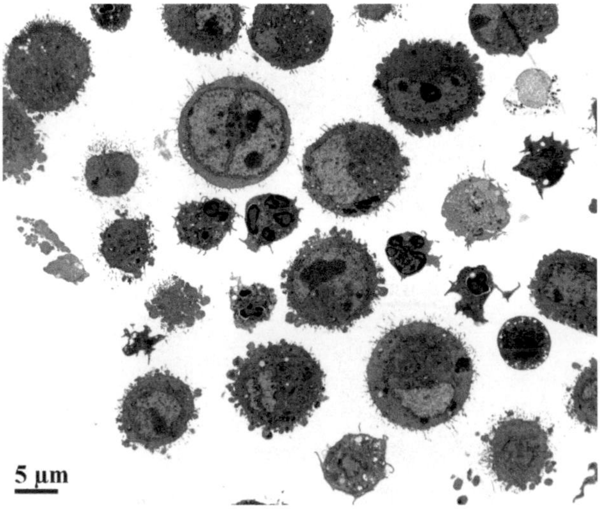

Fig. 5.20 Electron micrograph of poorly differentiated adenocarcinoma cells (TM, 2500×)
Poorly differentiated adenocarcinoma cells are numerous and scattered, rarely adhering into clusters

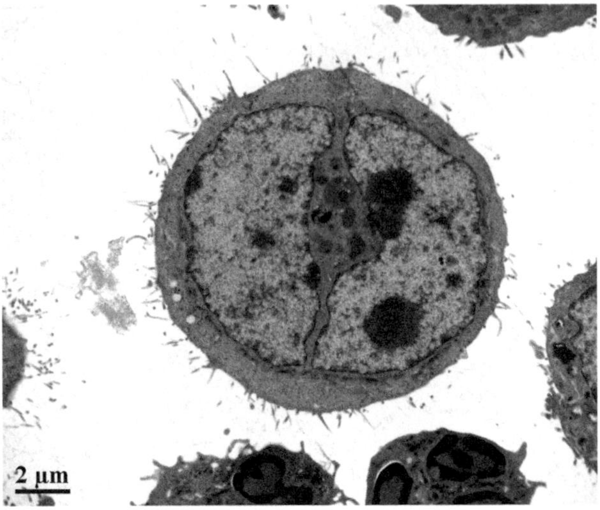

Fig. 5.21 Single poorly differentiated adenocarcinoma cell (TM, 8000×)
Nucleus cleaved into two parts, predominantly euchromatin, nucleolus large and well defined; few organelles, predominantly mitochondria, more free ribosomes and glycogen, cell body darkly stained; microvilli short and dense

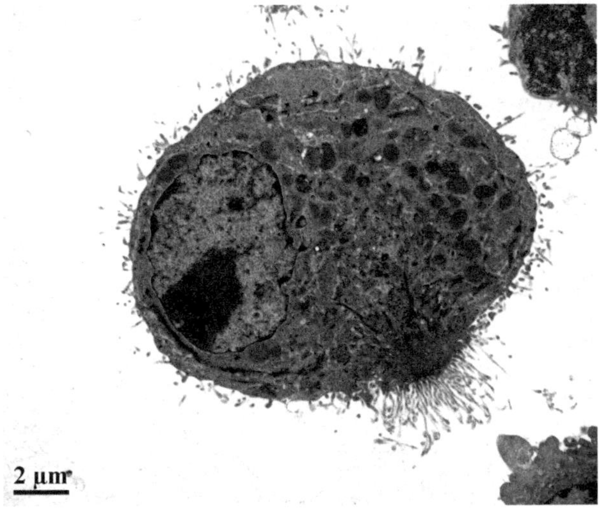

Fig. 5.22 Poorly differentiated adenocarcinoma cells with polarized microvilli, which are very "aggressive" and highly adhesive (TM, 8000×)

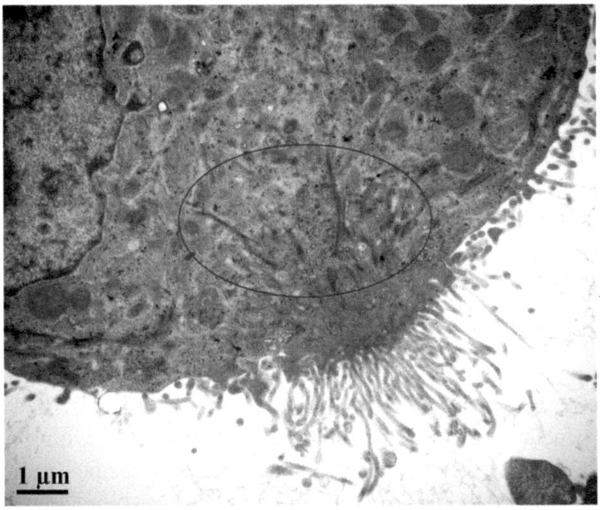

Fig. 5.23 A higher-magnification view of the area outlined in Fig. 5.22, showing bundles of tonofibrils and microfilaments (indicated by red circles) (TM, 15,000×)
Under the electron microscope, the tumor cells have obvious nuclear malformations, large nucleocytoplasmic ratio (poorly differentiated tumor), large nucleoli, and the chromatin of the nucleus is predominantly euchromatin. The cytoplasm was rich in organelles, with abundant free ribosomes (making the cells grayish-black after staining) and abundant glandular follicles (the source of adenocarcinoma). There are dense microvilli outside the cell membrane (pink under the light microscope), with bundles of microvilli on one side, and a large number of tonofibrils and microfilaments in the cytoplasm connected with them (red circle), which can be seen that the bundles of microvilli and the tight junction of the cells are closely related to each other, and it is a channel for the establishment of connection between the cells and other cells or tissues. The patient's extensive systemic metastasis and the very strong metastatic ability of the tumor cells may be closely related to the well-developed microvilli of such cells, which are conducive to implantation. In addition, when the tumor cells are small in size, they can easily pass through the barrier between cells or tissues and metastasize through blood or body fluids, which is also the reason for their extensive metastasis, and thus the prognosis is extremely poor

5.1.4 Poorly Differentiated Carcinoma of the Esophagus and Gastric Fundus

Case: Patient, elderly male, 65 years old, chronic course. Three months ago, he had abdominal pain without any cause, black stools, no fever, chills, nausea, vomiting, and other discomforts and went to a traditional Chinese medicine hospital for consultation. The results of whole-abdomen plain scanning + enhancement + CTA suggested that cancer of the gastric cardia was considered to be accompanied with multiple lymph node metastases in the surrounding area. The results of gastroscopy suggested cancer of the lower esophagus, cardia, and gastric fundus; and a continuous huge irregularly elevated ulcer was seen in the lower esophagus, cardia, and gastric fundus, and the bleeding was obvious, which was stopped after repeated spraying of prothrombin. Pathology report suggested low differentiated carcinoma of esophagus and low differentiated carcinoma of gastric fundus. On January 13, 2021, the Department of Pathology considered the following: esophageal squamous metaplasia and a small number of heterogeneous cells, suspected to be adenocarcinoma and poorly differentiated carcinoma of the gastric fundus. Immunohistochemical staining showed MSH6 (+), PMS2 (+), Ki-67 (+80%), CAM5.2 (+), HER-2 (1+), CD56 (-), SYN (-), CgA (-), P63 (-), and CEA (+). PET-CT report suggested that the wall of the lower esophagus and gastric cardia was abnormally thickened, and FDG metabolism was abnormally increased, so it was considered that the gastric cancer had involved the esophagus and gastric serosa layer, and there were multiple lymph node metastases in the parietal of the gastric cardia and the hepatogastric hiatus. Multiple small lymph nodes in the abdominal cavity, retroperitoneum, and bilateral diaphragmatic hiatus with mildly increased FDG metabolism are considered as inflammatory lymph nodes, and it is recommended to closely follow up the patient to exclude the possibility of individual metastasis. Neoadjuvant therapy is considered to be effective. The patient has a little fatigue, slightly reduced appetite, no fever, chills, bloody stools, black stools, and other discomforts and is to be admitted to the hospital for further treatment of "gastric space-occupying lesions." Cytomorphological examination of the pleural effusion fluid is presented in Fig. 5.24, 5.25, 5.26, 5.27, 5.28, 5.29, and 5.30.

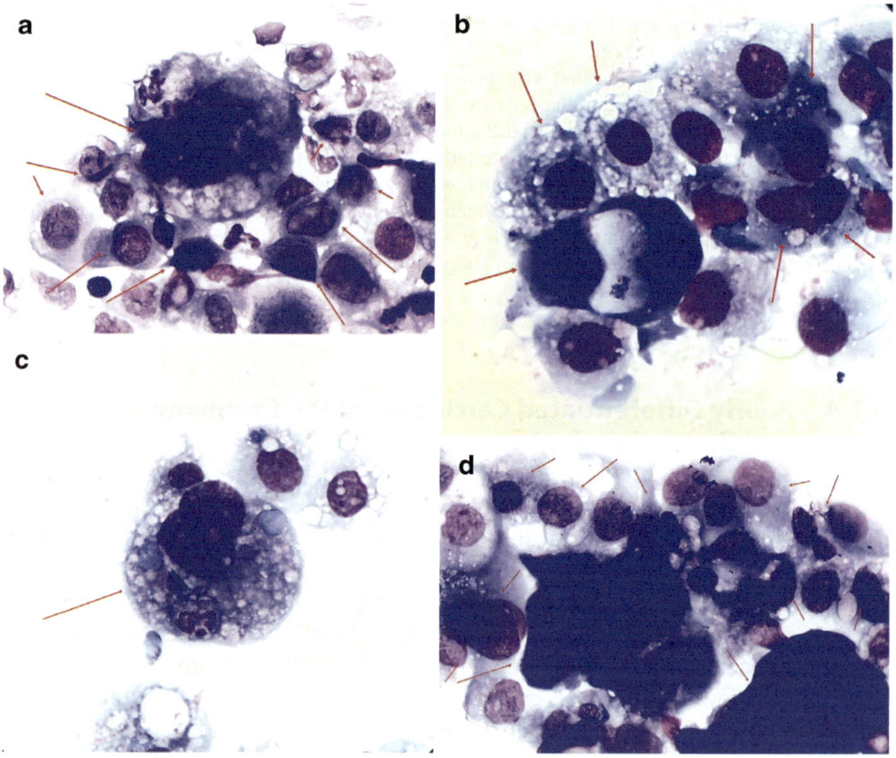

Fig. 5.24 Tumor cells of varying sizes (Ritter-Giemsa stain, 1000×)
Tumor cells are of different sizes, with the exception of a few tumor cells that are large in size, all other tumor cells are small in size, and are very easy to distinguish from mesothelial cells

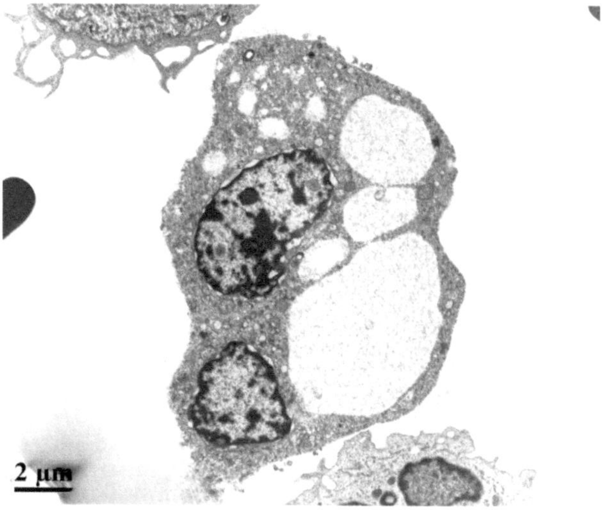

Fig. 5.25 Some of the better differentiated adenocarcinoma cells with more typical features (TM, 7000×)

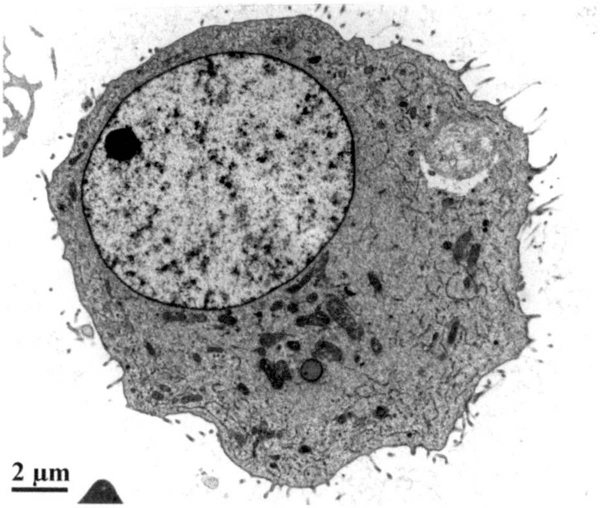

Fig. 5.26 Partially poorly differentiated adenocarcinoma cells (TM, 8000×)
The nucleus is dominated by euchromatin, large nucleoli; few organelles in the cytoplasm, light coloration; and short, dense microvilli outside the cell membrane

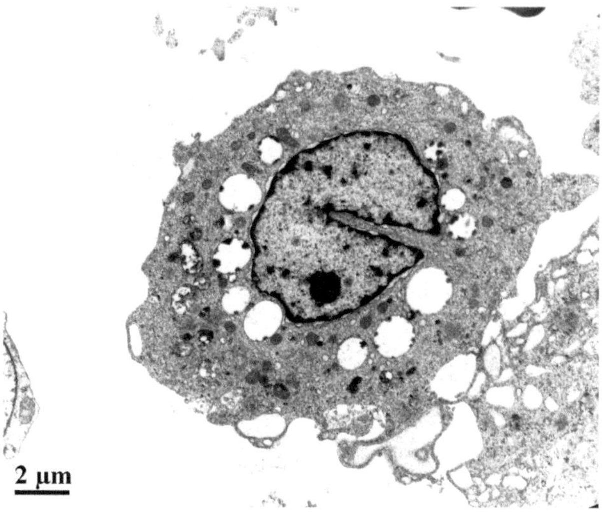

Fig. 5.27 Activated macrophages (TM, 8000×)
Usually, in addition to tumor cells, a large number of normal, activated, and nuclear heterogeneous cells, such as activated macrophages, are seen in the specimen; the cells in the figure have phagocytosis vesicles of different sizes, abundant primary and secondary lysosomes, long and sparse pseudopods, folded nuclei, and clearly defined nucleoli upon activation

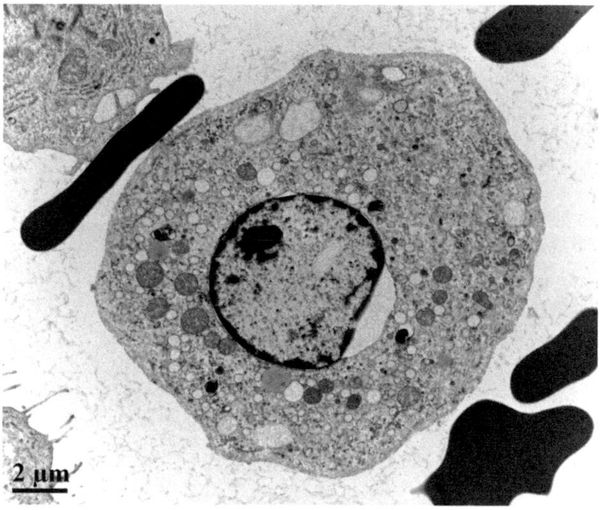

Fig. 5.28 Degenerate mesothelial cells (TM, 8000×)
The cell is in the apoptotic phase, with a rounded nucleus and obvious perinuclear clefts; the nucleus has loose chromatin and obvious gaps; the cytoplasm is generally swollen and fragmented in the organelles, and the pseudopods are not obvious

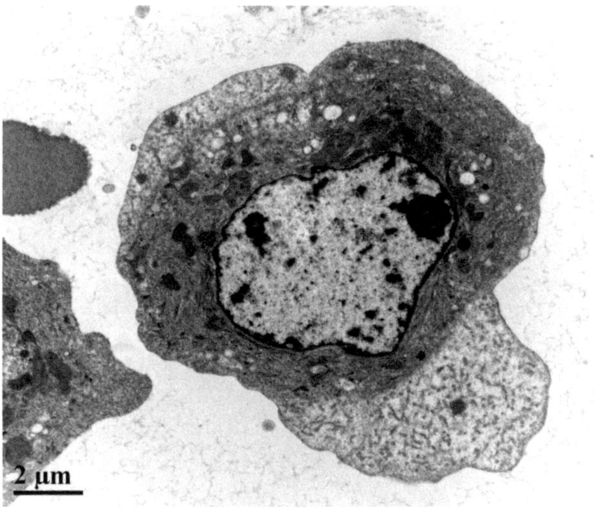

Fig. 5.29 Nuclear heterogeneous mesothelial cells (TM, 10,000×)
Key points for differentiating dysplastic mesothelial cells from poorly differentiated tumor cells: Poorly differentiated tumor cells are challenging to distinguish due to their low degree of differentiation. They typically exhibit fewer organelles, scant free ribosomes, pale staining, and very few tonofibrils. Their nuclei often appear immature, contributing to the diagnostic difficulty. In contrast, dysplastic mesothelial cells are well-differentiated, representing activated mesothelial cells in response to stimulation. They display markedly more organelles, abundant free ribosomes, and cell pseudopods that frequently form wart-like projections—distinct from the microvilli of tumor cells. A notable increase in tonofibrils is often observed, suggesting a possible epithelial origin and strong resistance to external factors.Unlike in histopathological specimens, cytomorphology in body fluids predominantly reveals isolated single cells, with only occasional small clusters; therefore, intercellular junctions commonly seen in tissue sections are rarely observed. This morphological feature has also been described in cases of invasive lobular carcinoma of the breast (see above)

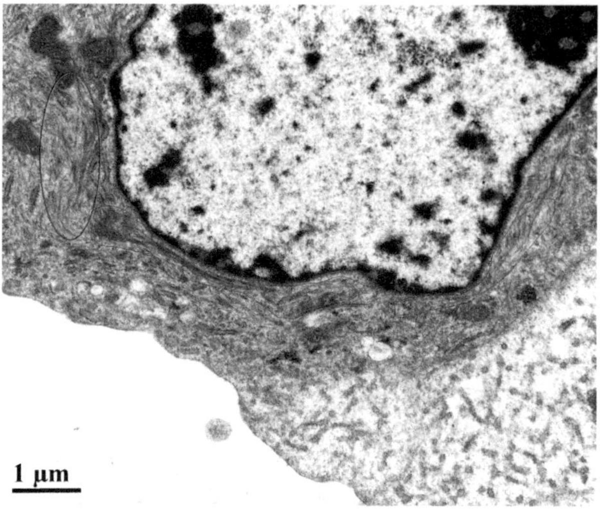

Fig. 5.30 Localized magnified electron micrographs of nuclear heterogeneous mesothelial cells (TM, 20,000×)
Local magnification of a nuclear heterogeneous mesothelial cell with bundles of tonofibrils in the red circle. Intracellular microtubules, microfilaments, and tonofibrils are well developed. Some cytoplasmic interstitial spaces are increased and organelles are swollen, suggesting a tendency toward apoptosis

5.1.5 Infiltrating Poorly Differentiated Carcinoma of the Breast

Case: Patient, female, 41 years old, with chronic disease. diagnosed with breast cancer 3 years ago, nausea with abdominal pain for half a month. She was diagnosed 3 years ago and underwent mastectomy. Pathologic findings suggested invasive poorly differentiated carcinoma. The results of multiple chemotherapy were unsatisfactory. On follow-up examination, there were enlarged axillary lymph nodes, enlarged right supraclavicular lymph nodes, and invasive adenocarcinoma cells were found by puncture, with ER (-), PR (-), weak CerbB-2 (+), and Ki-67 (+) >90%. The morphologic examination of pleural fluid is shown in Figs. 5.31, 5.32, 5.33, 5.34, 5.35, 5.36, 5.37, and 5.38.

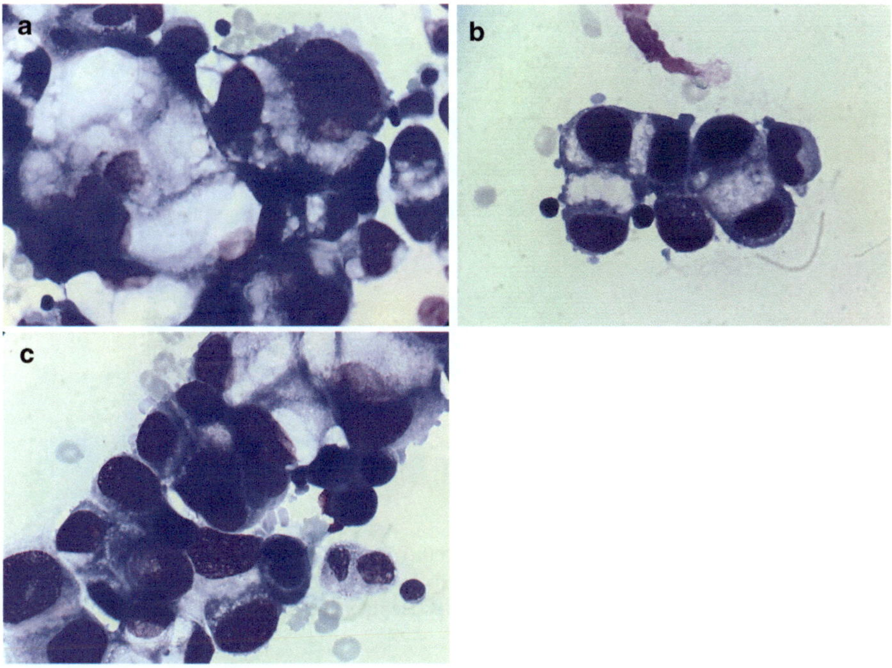

Fig. 5.31 Poorly differentiated adenocarcinoma nest (Ritter-Giemsa stain, 1000×)
The patient was pathologically diagnosed as an invasive, poorly differentiated adenocarcinoma, and the adenocarcinoma cells were seen to be poorly differentiated and distributed in clusters. Occasionally, adenocarcinoma cells with a relatively good degree of differentiation were seen; mostly poorly differentiated tumor cells with a large nucleo-cytoplasmic ratio, deep staining of the cells, loose nuclear chromatin, low cytoplasmic volume, small or absent acinus, and irregular cell bodies were seen

Fig. 5.32 Adenocarcinoma cells (TM, 2500×)

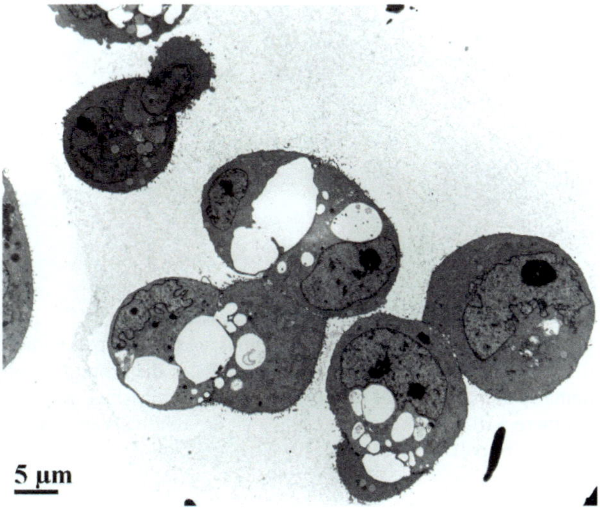

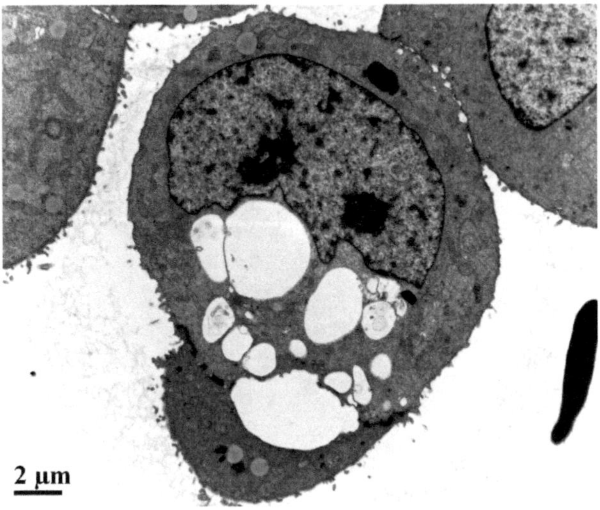

Fig. 5.33 Single adenocarcinoma cells (TM, 7000×)
Nucleus is malformed, acinus is huge, cell body is irregular, abundant organelles, free ribosomes and glycogen are densely packed, the whole cell is deeply stained

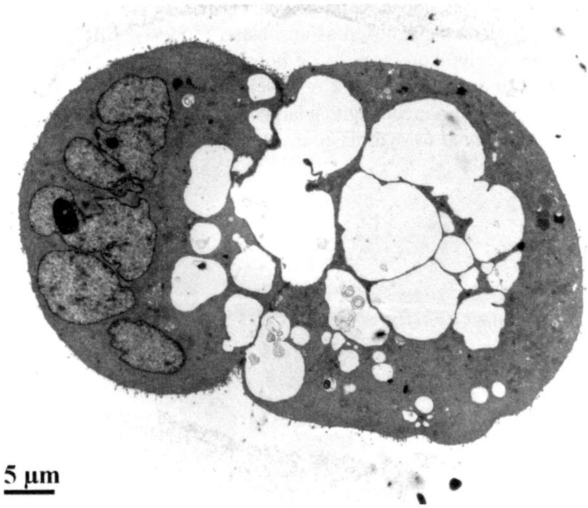

Fig. 5.34 Bulky adenocarcinoma cells (TM, 3000×)
Usually, poorly differentiated carcinomas are not so large. Pathologically, the degree of differentiation is based on the proportion of such cells. If the proportion of poorly differentiated adenocarcinoma cells in the cancer nest is high, then it is classified as poorly differentiated adenocarcinoma, but there will always be some cells that are relatively highly differentiated, not absolutely purely one type of cell

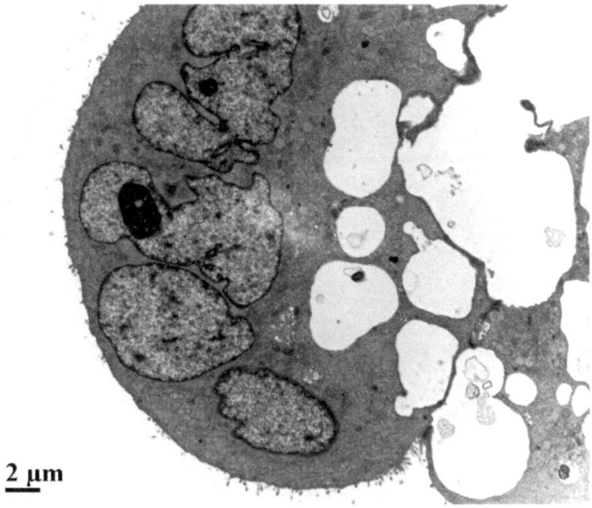

Fig. 5.35 Localized magnification of adenocarcinoma cells (TM, 5000×)
Tumor cells have fragmented nuclei into multiple pieces, acini are large and numerous, crowding the nucleus into an off-set position, microvilli on the cell membrane are short and dense, free ribosomes and glycogen are abundant, and the cells are deeply stained

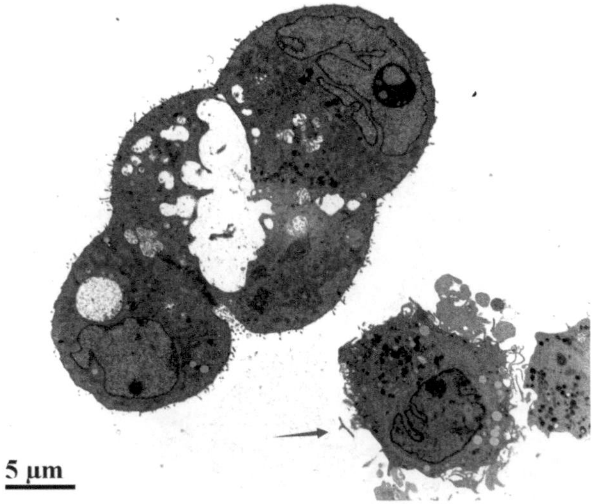

Fig. 5.36 Adenocarcinoma cells (TM, 4000×)
The nucleus is obviously deformed, the acinus is huge, and the nucleo-cytoplasmic ratio of poorly differentiated tumor cells is large. The arrows point to activated macrophages with folded and twisted nuclei, visible obvious nucleoli, abundant phagocytic vesicles and lysosomes in the cytoplasm, and slender and curled outer pseudopods of the cytoplasm, which are obviously different from the fine, short and dense microvilli of the tumor cells

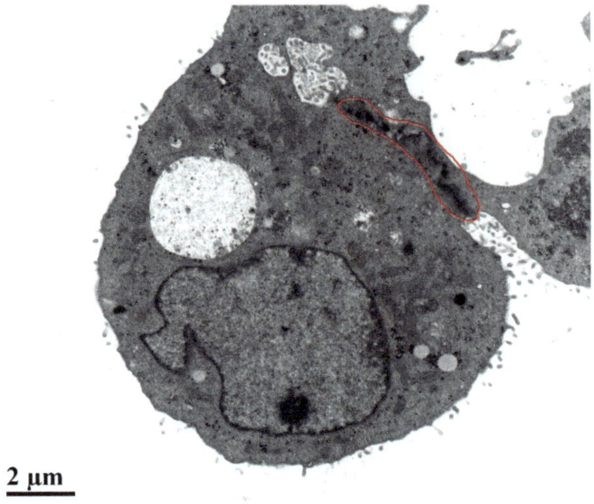

Fig. 5.37 Localized magnification of adenocarcinoma cells (TM, 10000×)
The mitochondria of this tumor cell are abundant, and free ribosomes appear scattered or in clusters, suggesting that the cell synthesis and proliferation are exuberant. Intercellular junctional complexes are visible (outlined in red)

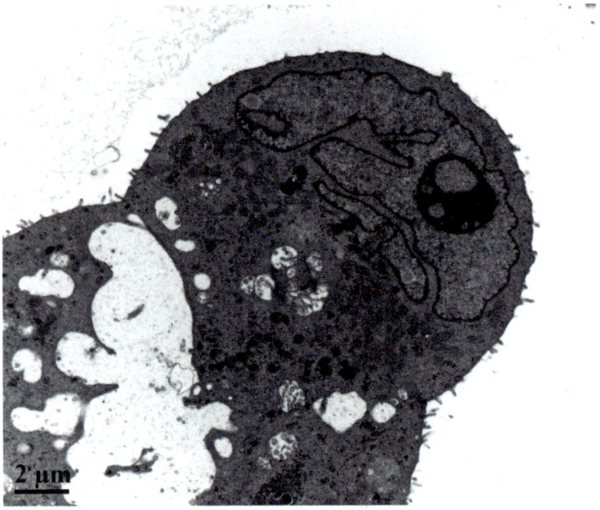

Fig. 5.38 Poorly differentiated adenocarcinoma cells (TM, 8000×)
The tumor cells had huge nucleoli, malformed and fragmented nuclei, abnormally abundant mitochondria in the cytoplasm, and glycogen appeared in clusters, suggesting that the tumor cells were actively proliferating. And the tumor cell invasion and metastasis ability is strong, the prognosis is poor; the patient was admitted to the hospital, the disease progressed rapidly, and the patient died shortly afterward

5.1.6 Gastric Signet-Ring Cell Carcinoma, Poorly Differentiated Adenocarcinoma

Case: The patient is an elderly woman, 73 years old, with a chronic course of disease, who underwent laparoscopic radical gastrectomy + Roux-Y anastomosis in 2019, which was uneventful, and she recovered well after surgery. The pathology report of the postoperative radical resection of the whole gastric tumor indicated that ① the "lesser curvature of stomach" was a poorly differentiated adenocarcinoma, part of which was gastric signet-ring cell carcinoma, infiltrating and ulcerating type (Lauren classification: diffuse type), with a size of 8cm×7cm; the tumor infiltrated into the gastric serosa layer and invaded the submucosal, muscular, and epimysial layers of the esophagus, the nerve involved (+), the vascular involved (+); (2) the esophageal margins and self-examination of the lower margins were negative; (3) self-examination of lymph nodes in the greater curvature of stomach (5/14) and lesser curvature of stomach (3/3), and metastatic cancer in lymph nodes of group VIII (0/3), IX (0/1), X (0/1), and XI (0/2) were sent for examination. The results of immunohistochemistry were Ki-67 (+70%), CAM5.2 (+), HER-2 (0), E-cadherin (+), EGFR (+), CK (Pan) (+), CK20 (partially +), CK7 (+), and P53 (-). PET-CT results suggested that after laparoscopic radical gastrectomy + Roux-Y anastomosis, the wall of jejunum and esophagus end-to-side anastomosis was thickened and FDG metabolism was mildly increased, while the wall of intestine side anastomosis did not show significant thickening and FDG metabolism was mildly increased, and a little inflammatory lesion of the anastomosis was considered. Inclusion of fluid in the uterorectal recess with thickening of the adjacent peritoneum and mildly elevated FDG metabolism is considered to be due to implantation metastasis; there is a small amount of perihepatic fluid, and there is intestinal obstruction. Diagnosis of intestinal obstruction with abdominal implantation metastasis was made, and jejunum side-to-side anastomosis + jejunum nutritional fistula + intestinal adhesion release + abdominal infusion port placement was performed, and the postoperative recovery was good. Now, she is seeking for review and chemotherapy again and intends to be admitted to the hospital for further treatment as "postoperative chemotherapy for gastric malignant tumor." The cytomorphologic examination of body fluids is shown in Figs. 5.39, 5.40, 5.41, 5.42, 5.43, and 5.44.

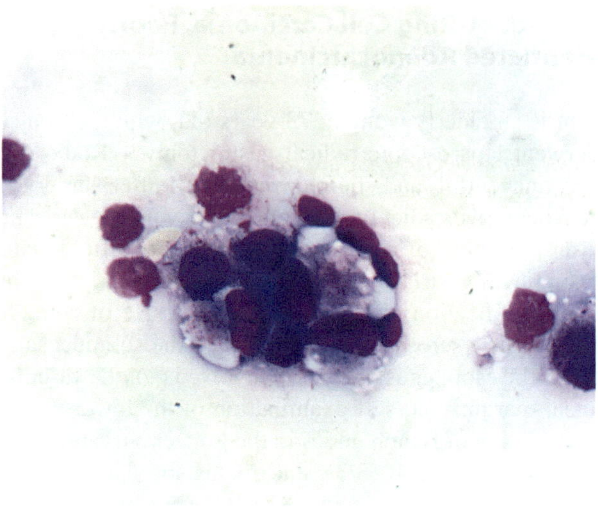

Fig. 5.39 Poorly differentiated adenocarcinoma cells (Ritter-Giemsa stain, 1000×)
Poorly differentiated adenocarcinoma cells appear in clusters and are more difficult to distinguish from nuclear heterogeneous mesothelial cells under light microscopy

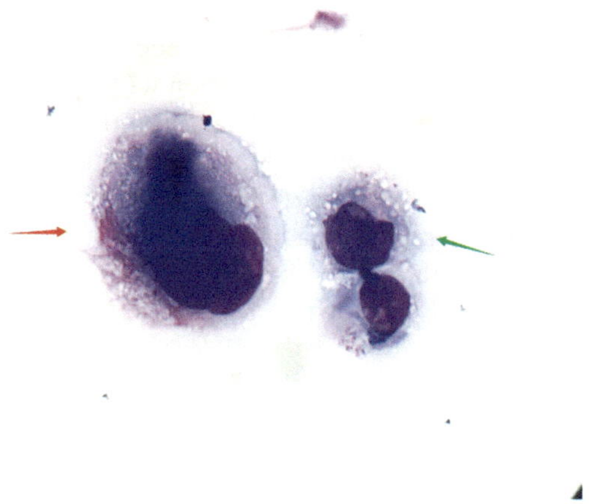

Fig. 5.40 Gastric signet-ring cell carcinoma cells (red arrows) versus poorly differentiated adenocarcinoma cells (green arrows) (Ritter-Giemsa stain, 1000×)
Gastric signet-ring cell carcinoma cells and poorly differentiated adenocarcinoma cells have small cell bodies, malformed nuclei, and small acini in the cytoplasm. Gastric signo-cytoplasmic-ring cell carcinoma cells have relatively high cytoplasmic volume and eccentrically located nuclei and are also poorly differentiated tumors; poorly differentiated adenocarcinoma cells have a large nucleo-cytoplasmic ratio and low cytoplasmic volume

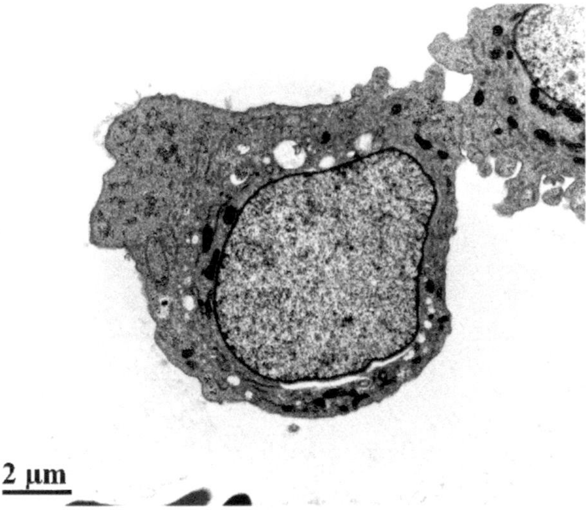

Fig. 5.41 Poorly differentiated adenocarcinoma cells (TM, 10 000×)
Tumor cells have a large nucleo-cytoplasmic ratio, nuclei dominated by euchromatin, few organelles in the cytoplasm, mitochondria predominate, glycogen appears scattered or in clusters and is black, free ribosomes are densely gray, and the acini are small and few in number, with perinuclear fissures, suggesting degeneration of the cell, proximity to fragmentation, apoptosis

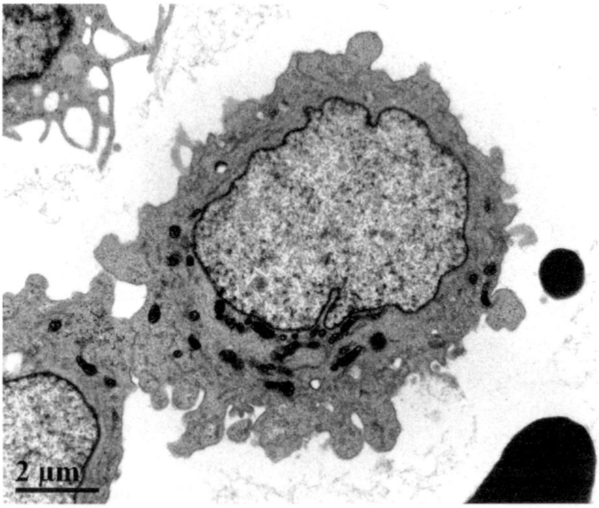

Fig. 5.42 Poorly differentiated adenocarcinoma cells (TM, 8000×)
Tightly connected to cells in Fig. 3.174, both are poorly differentiated adenocarcinoma cells with consistent characteristics and more pseudopodia

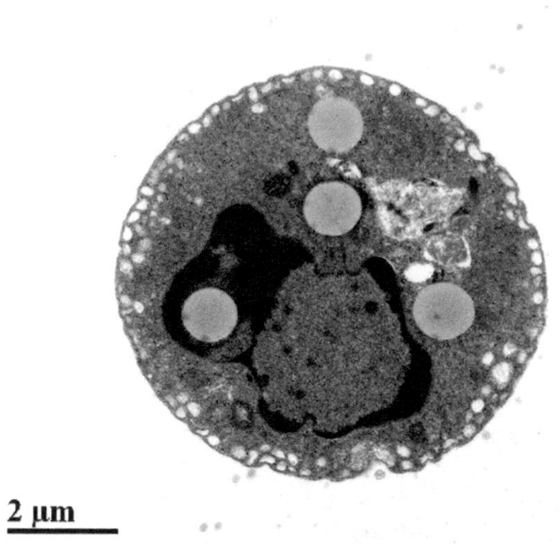

Fig. 5.43 Gastric signet-ring cell carcinoma cells (TM, 10,000×)
Nuclear malformation, deviation, large cell body, abundant acini, and grayish black cell body

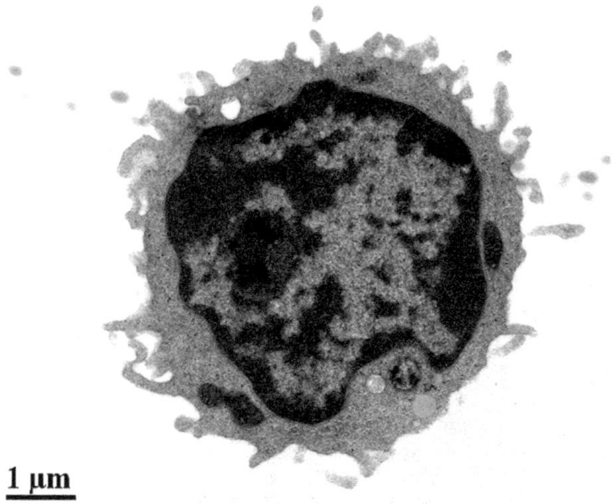

Fig. 5.44 Dormant lymphocytes (TM, 12000×)
Lymphocytes and macrophages are common cells in bodily fluids. Dormant lymphocytes have a larger nucleo-cytoplasmic ratio, which is obviously smaller than tumor cells. The nuclear chromatin of lymphocytes is mainly heterochromatin, which aggregates into lumps, and there are few organelles in cytoplasm, and a few mitochondria are easily seen. Cell microvilli are thick and short, usually thicker than tumor cell microvilli

Light and Electron Microscopy Comparison of Other Abnormal Humoral Cells

6

Dayong Gu and Ying Wang

6.1 Lymphoblastic Leukemia/Lymphoma

Lymphoma is a malignant tumor originated from lymphohematopoietic system, which mainly manifests as painless enlarged lymph nodes, and enlarged liver and spleen, and all tissues and organs of the whole body can be involved, accompanied by systemic symptoms such as fever, night sweating, emaciation, itching, etc. Lymphoma of the chest is very easy to metastasize and grow in the chest cavity, especially close to the pleura. Since the thymus gland is located in the upper part of mediastinum behind the sternal stalk and it is an important developing and maturing organ of the lymphatic system, the lymphoma in the chest is very easy to be metastasized and implanted in the thoracic cavity, especially the tumors of tissues and organs close to the pleura. It is very common to find lymphoma cells in pleural fluid, and rapidly progressive and aggressive lymphomas are particularly common. Acute lymphoblastic leukemia is a tumor of precursor lymphocytes (lymphoblasts), with a predominance of invasion of the bone marrow and peripheral blood (usually more than 25%) and usually also of the lymph nodes. Abnormal lymphocytes can be seen in pleural fluid, ascites, and cerebrospinal fluid. After treatment, many residual abnormal lymphocytes often hide in these cavities that are not easily penetrated by medications, resulting in recurrence and metastasis of the tumor.

D. Gu (✉) · Y. Wang
Laboratory Department, Shenzhen Second People's Hospital, Shenzhen, Guangdong, China

D. Gu, Y. Wang (eds.), *Morphological Diagnosis of Humoral Cells under Light and Electron Microscopy*, https://doi.org/10.1007/978-981-95-8668-4_6

6.1.1 Extensive Metastasis of Lymphoma

Case: Patient, female, 73 years old, with chronic disease. One week ago, she was admitted to the hospital with "abdominal distension with cessation of defecation for 2 days." The CT examination report of abdomen and pelvis suggested that occupying lesion in the distal part of descending colon, malignant tumor possible; multiple nodules and masses were seen in the abdominopelvic cavity, inguinal cavity, peritoneum, greater omentum, mesentery, retroperitoneum, angle of heart diaphragm, thoracic paraspinal region, and mediastinum, metastasis possible. Enhanced CT examination is recommended. There were bilateral thickening of adrenal glands with localized suspicious nodules, slight thickening of gallbladder wall, and a small amount of pelvic fluid. Attachment: Bilateral pleural cavity with small amount of effusion. The patient presented with chest tightness and shortness of breath, inability to lie down, low breath sounds in both lungs, and swelling of both lower extremities and was referred to the Department of Respiratory and Critical Care Medicine for consultation and recommended referral for further treatment. The patient is now in good spirits, with decreased physical strength, poor appetite, fair sleep, no significant change in body weight, no bowel movement for 1 week, and normal urination. The patient was in the terminal stage and died 3 days after admission. The cytomorphology of the pleural effusion is shown in Figs. 6.1, 6.2, 6.3, 6.4, and 6.5.

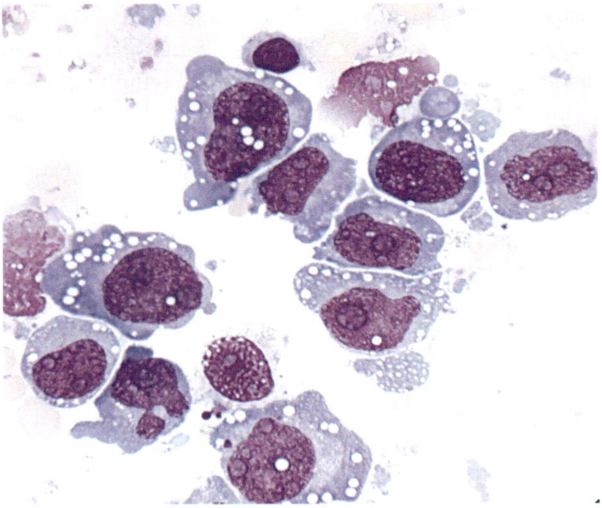

Fig. 6.1 Lymphoma cells (Ritter-Giemsa Stain, 1000×)
Vacuolar degeneration is evident, nucleoli of the nuclei are huge, smudge cells are increased

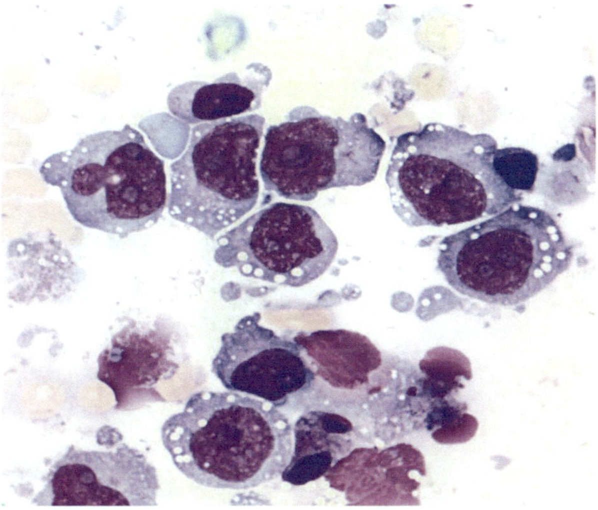

Fig. 6.2 Lymphoma cells (Ritter-Giemsa Stain, 1000×)
Nuclear malformations are seen, cytoplasmic plastids (formed by shedding of cytoplasm; there is no nucleus, part of the cytoplasm; granules are visible) are prevalent, and the increase in smudge cells suggests that the cells are susceptible to apoptotic rupture

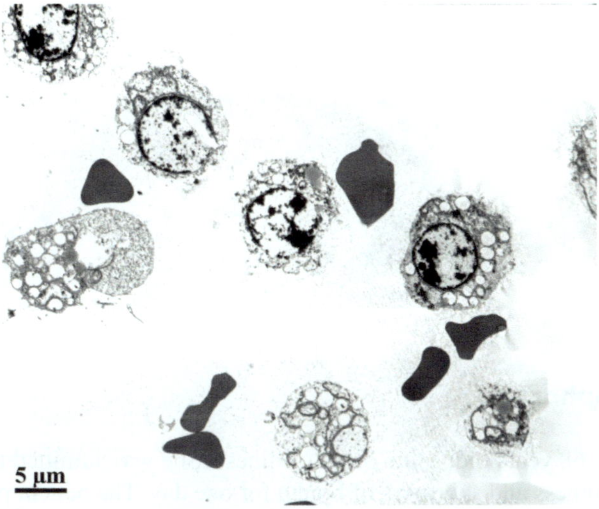

Fig. 6.3 Lymphoma cells and cytoplasmic plastids (TM, 3000×)
The chromatin of the cell nucleus was obviously loose, with obvious whitening, and the cell membrane was broken and incomplete, suggesting that the cells were in the pre-apoptotic stage. Because the lymphocytes were immersed in pleural water for a long time, and the abnormal lymphocytes were easy to be broken and apoptotic, forming smudge cells, so the cells were mostly broken and vacuolated obviously under the electron microscope

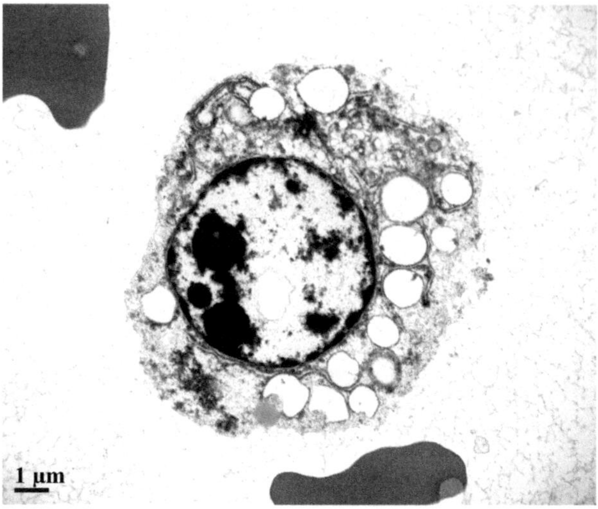

Fig. 6.4 Lymphoma cells (TM, 10,000×)
The nucleolus is large and obvious; there are many vacuoles in the cytoplasm, the endoplasmic reticulum is swollen and degenerated, and the integrity of the cell membrane is poor, suggesting that the cell is in the pre-apoptotic fragmentation stage

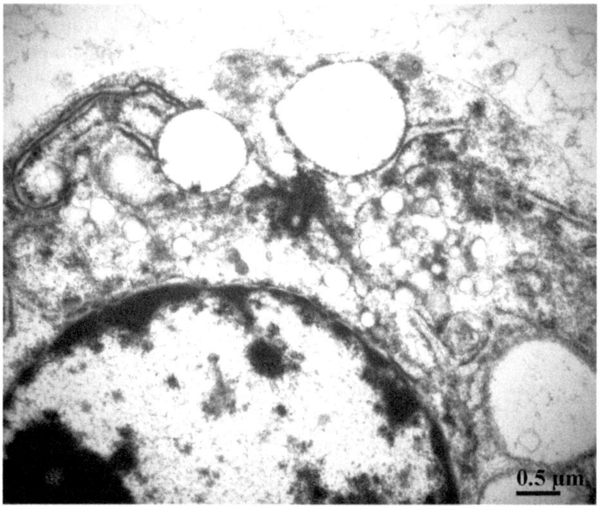

Fig. 6.5 Localized magnified electron micrograph of lymphoma cells (TM, 40,000×). All organelles were swollen and lysed, and vacuoles were obvious

6.1.2 Lymphomas

Case: Female, 68 years old, acute onset of illness. She was admitted to the hospital with chest tightness and shortness of breath for one day. The patient presented with chest tightness and shortness of breath 1 day ago without any obvious triggers, so she was admitted to the emergency room. Chest CT showed that there was a large amount of pleural effusion on the right side of the chest, insufficient expansion of the right lung, and a small amount of effusion on the left side of the chest. A chest catheter was placed to drain the pleural effusion, and the routine examination of the pleural fluid showed that the nucleated cell count was 13,000/μl, with 85% of abnormal cells, and the possibility of lymphoma cells was considered. For further treatment, the patient was admitted to the hospital as "pleural effusion to be investigated." The cytomorphologic examination of the pleural fluid was as follows, as shown in Figs. 6.6, 6.7, 6.8, and 6.9.

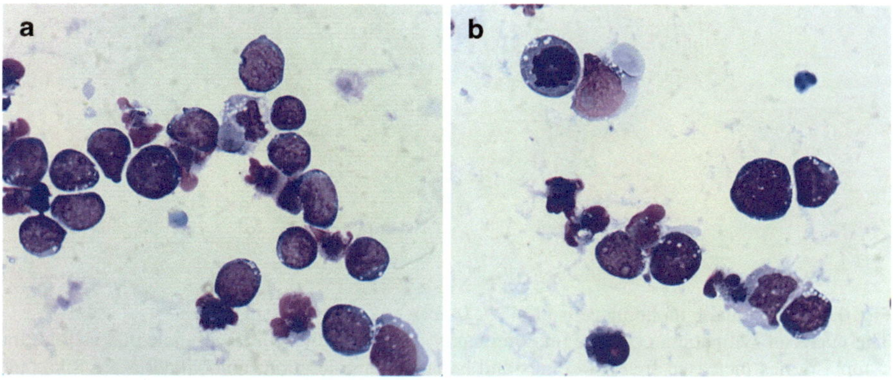

Fig. 6.6 Lymphoma cells (Ritter-Giemsa Stain, 1000×)
Large nucleo-cytoplasmic ratio, fine nuclear chromatin, large and conspicuous nucleolus; low cytoplasmic volume, with vacuoles; increased smudge cells; easy to see abnormal mitotic figures

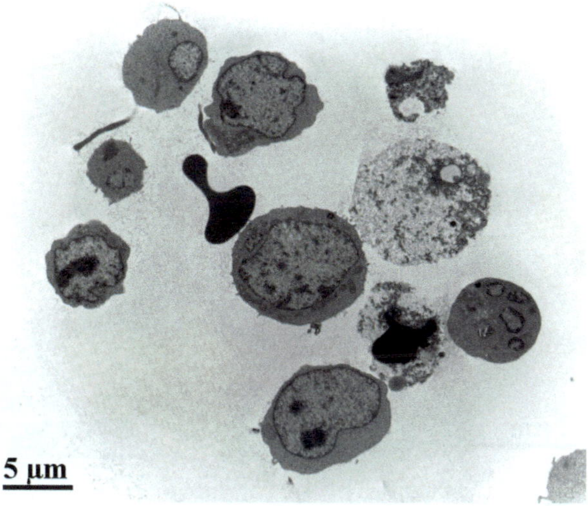

Fig. 6.7 Lymphoma cells (TM, 4000×) fewer vacuoles in cytoplasm, more complete cells, smudge cells easily seen

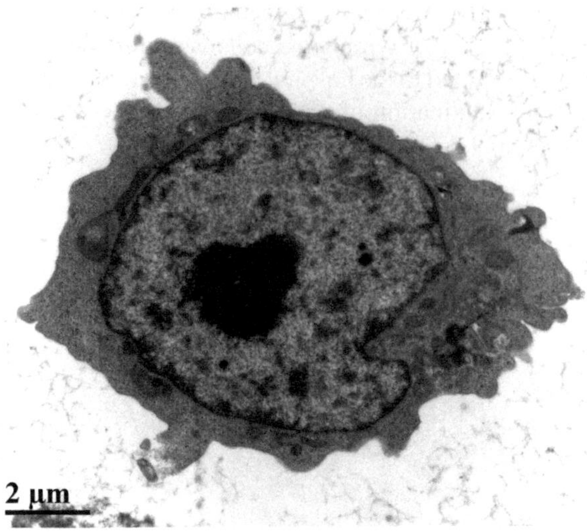

Fig. 6.8 Individual lymphoma cells (TM, 12,000×)
The nuclei of lymphoma cells are predominantly euchromatin, with large nucleolus visible; the cytoplasm has increased free ribosomes and is darker in coloration; mitochondria are thick and markedly increased; the cell bodies are slightly larger than those of dormant lymphocytes, and the cell size is similar to that of reactive lymphocytes, which can be compared with the next case

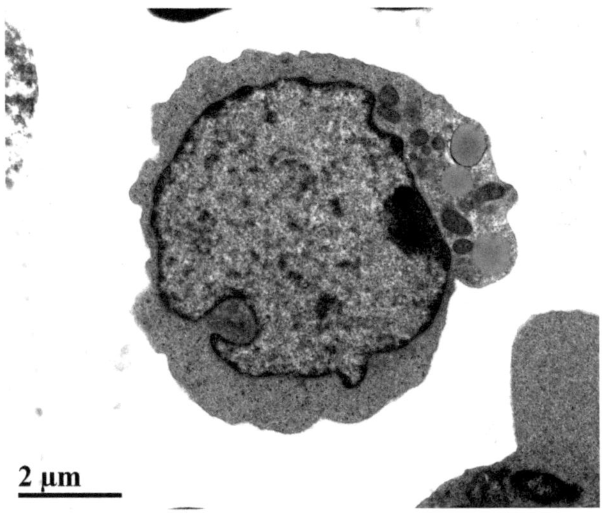

Fig. 6.9 Lymphoma cells (TM, 15,000×)
Secretory vesicles are seen in the cytoplasm of some lymphoma cells, which are gray, homogeneous, and slightly different from the vacuoles formed by secondary lysosomes and may be associated with the secretion of clonal immunoglobulins or other abnormal proteins

6.2 Reactive Lymphocyte

Case: Patient, male, 23 years old, acute course. Ten days ago, he developed fever with no obvious cause, and the highest temperature was 39.8 °C, accompanied by chills, headache, fatigue, dry cough, sore throat, and right-sided chest and back pain in the morning, which was related to the posture and breathing and lasted for tens of minutes and then relieved. Leukocyte count was 13.12×10^9/L, ultrasensitive CRP 17.44 mg/L, and chest CT showed a little inflammation in the right lung and thickening of the right pleura. Cefotaxime was given to fight infection for 3 days, but the symptoms did not show significant improvement, and the temperature was still around 38 °C repeatedly, and then moxifloxacin injection was changed to intravenous drip treatment, the symptoms were relieved, and the temperature decreased. Two days ago, the patient's temperature rose again, and the symptoms were the same as before. The blood test showed a decrease in leukocytes, but the CRP of 120 mg/L was higher than before. Three days later, a CT examination of the chest showed scattered inflammation in both lungs (mainly in the lower lobe of the right lung); pleural effusion on the right side, accompanied by partial insufficient expansion of the lower lobe of the right lung, and encapsulated effusion in the pleura of the right oblique fissure; the patient was admitted to the hospital as "pneumonia" in order to seek further treatment. After admission, the patient's spirit was fine, physical strength was normal, appetite was normal, sleep was normal, weight did not change significantly, bowel movements were normal, and urination was normal. Two days after admission, routine biochemical examination of pleural effusion showed: adenosine deaminase 55 U/L ↑; sputum bacterial + fungal culture sensitivity: a large number of gram-negative bacilli, a large number of gram-positive cocci. The chest tube drained 600 ml of yellowish fluid. Combined with the patient's adenosine deaminase 55 U/L ↑, the possibility of tuberculous pleurisy was considered to be high, and the pleural fluid was found to be positive for tuberculosis DNA, and the sputum was not found to be resistant to acid-filled bacilli. The cytomorphologic results of the body fluid are shown in Figs. 6.10, 6.11, and 6.12.

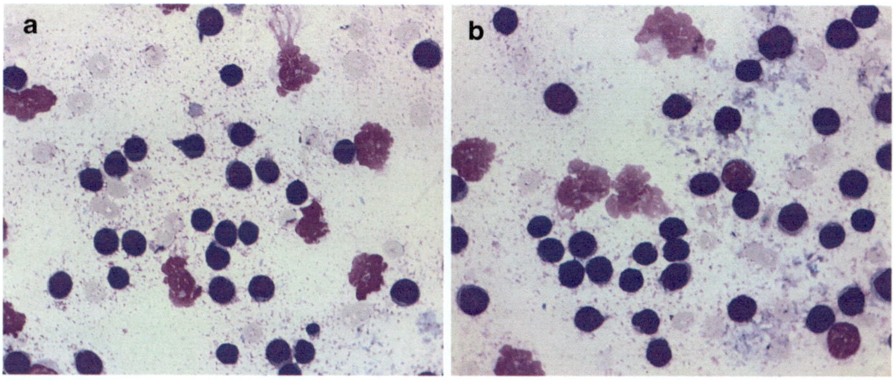

Fig. 6.10 Significantly increased lymphocytes (Ritter-Giemsa stain, 1000×). Reactive lymphocytes are seen; smudge cells are increased; large numbers of pink necrotic granules are seen in the background

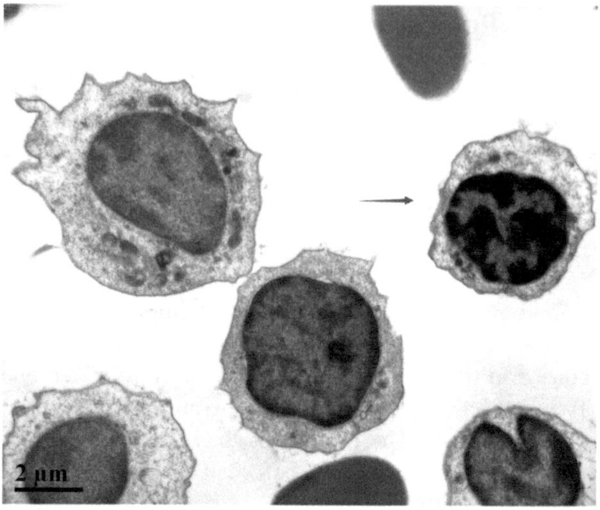

Fig. 6.11 Lymphocytes (TM, 10,000×)
Dormant lymphocytes (red arrows); the remaining lymphocytes are activated lymphocytes. The main difference is the change in nuclear chromatin, with heterochromatin predominating in dormant lymphocytes and euchromatin predominating in activated lymphocytes, with nucleolus visible

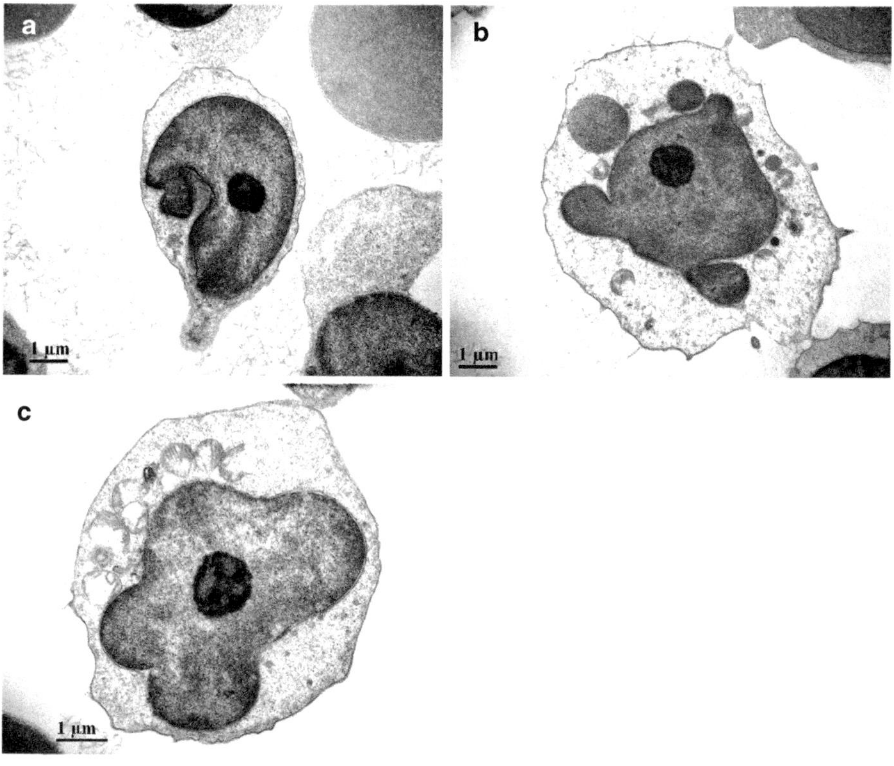

Fig. 6.12 Activated lymphocytes (**a**. TM, 15000×; **b**. TM, 15000×; **c**. TM, 20000×)

The nuclei of activated lymphocytes are variable and irregular, with predominantly euchromatin and obvious nucleolus; the cytoplasmic volume increases, and the number of organelles may increase. Lymphocytes are easily broken in body fluids, and the mitochondria in Figure C are swollen and broken, suggesting the beginning of apoptosis. For beginners, it is not easy to distinguish reactive lymphocytes from lymphoma cells morphologically, and flow cytometry and immunohistochemistry are needed to confirm the diagnosis.

6.3 Mesothelioma

Mesothelial cells are simple squamous epithelium that develop from mesoblasts and cover the surface of the pleura, peritoneum, and pericardium to form the mesothelium. Pleural mesothelioma is a primary tumor of the pleura, and there are two types of mesothelioma: confined (mostly benign) and diffuse (all malignant), with diffuse malignant mesothelioma being one of the worst prognostic tumors of the chest. The majority of patients are 40–70 years old, and there are more males than females. Peritoneal mesothelioma refers to tumors originated from peritoneal mesothelial cells, and its clinical manifestations are not characteristic, and common signs and symptoms include abdominal pain, ascites, abdominal distension, and abdominal mass. Peritoneal mesothelioma accounts for about 20% of all mesothelioma cases, with the age of onset ranging from 2 to 92 years old, with an average age of 54 years old, of which about 63% of cases are between 45 and 64 years old, and the disease is rare in children. Clinical mesothelioma patients are rare, and because the lesion is located in the mesothelium, cytomorphological examination of body fluids is often one of the means of direct detection of the tumor.

Case: Young patient, female, 27 years old, with chronic disease course. Two years ago, there was no obvious inducement of abdominal distension and discomfort, which was not related to eating and activities, accompanied by edema of both lower limbs, no fear of cold and fever, abdominal pain, diarrhea, chest tightness, shortness of breath, no gross hematuria, and other discomfort. She went to the emergency department of our hospital for treatment. After completing relevant examinations, she was considered as "renal anemia of type 1 diabetic nephropathy stage V" and was transferred to our department after treatment of hypoglycemia and diuresis. Abdominal distension improved significantly after hemodialysis, diuresis, albumin supplementation, and hypoglycemic treatment. Over the past 2 years, the patient has experienced recurrent abdominal distension. One week ago, the patient felt significant abdominal distension again, accompanied by coughing and no phlegm, conscious facial swelling, no obvious edema in both lower limbs, no fear of cold and fever, no abdominal pain, no obvious chest tightness, no chest pain, no shortness of breath, no nocturnal sitting breathing, etc. For the treatment of ascites, she was admitted to the department of our hospital in a state of "hemodialysis." The patient has been "diabetic" for more than 12 years and has been given "subcutaneous injection of aspartate insulin 8U early, 4U medium and 8U late combined 10U before bedtime" for a long time to control blood sugar. The patient has a history of "hypertension" for more than 2 years, with a maximum blood pressure of about 200/100mmHg. Currently, she has been taking 12.5mg of carvedilol orally once a day and 30mg of nifedipine twice a day for a long time, with moderate blood pressure control. And she has a personal history of elevated tumor markers and syphilis, without special treatment. In January 2019, long-term hemodialysis catheter implantation was performed. In March 2019, arteriovenous fistula with renal dialysis was performed. The patient has acceptable spirit, poor physical strength, poor appetite, mediocre sleep, significant weight gain, normal stool, and had no urine after being transferred to another department. Tumor markers (female) (serum): carbohydrate antigen 12588.5U/ml ↑, human epididymal protein 4 1461.0pmol/L ↑, and gastrin-releasing peptide precursor 258.8pg/ml ↑. In July 2021, the infectious disease inspection results were positive for hepatitis B surface antibody and syphilis specific antibody (TP Ab). Pathological liquid-based cytology examination of ascites: tumor cells were detected, indicating the possibility of malignant mesothelioma. Morphological examination of body fluid cells is shown in Figs. 6.13, 6.14, 6.15, 6.16, 6.17, 6.18, and 6.19.

Abundant mitochondria, inclusion bodies visible; phagocytic cell membrane boundaries (arrows) visible; tumor cell microvilli abundant, short, and dense

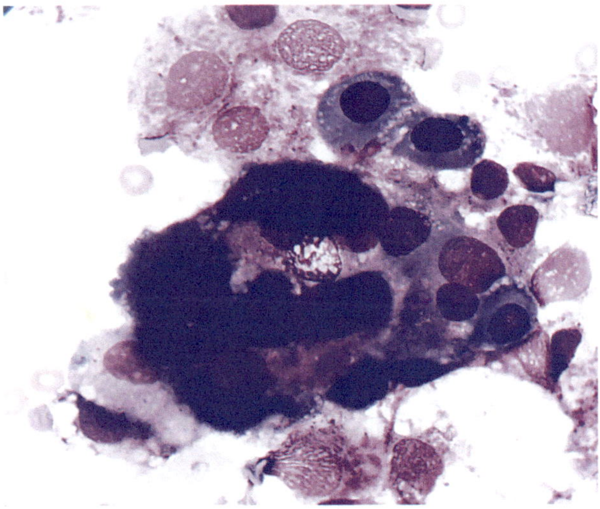

Fig. 6.13 Disordered mesothelioma cells (Ritter-Giemsa stain, 1000×)
Mesothelial cells often have mesothelial pores, and normal mesothelial pores are neatly aligned and uniform in size, with eosinophilic protein secretions stained pink in the center; in this patient, the mesothelial pores had lost their polarity of alignment and the cells were of different sizes

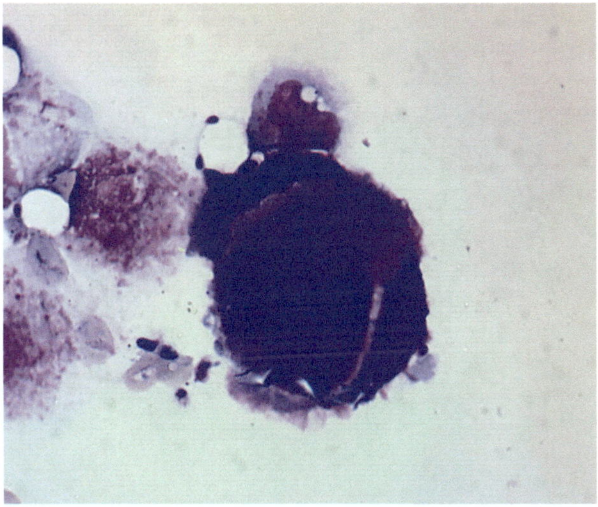

Fig. 6.14 Carcinoma nests of a malignant mesothelioma (Ritter-Giemsa stain, 1000×)
Abnormally arranged aberrant mesothelioma cells with distinct nucleolus visible

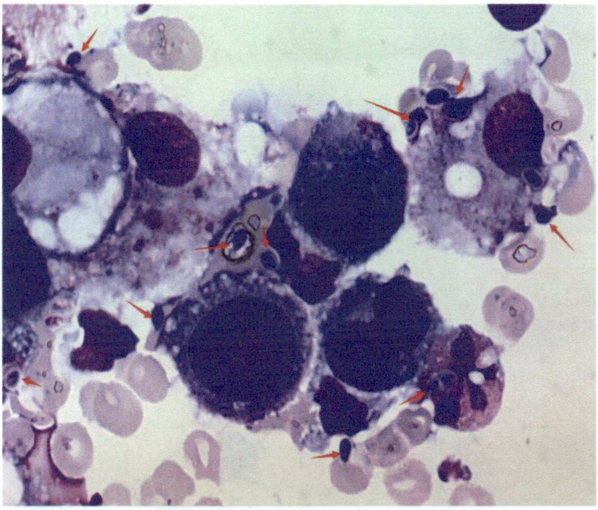

Fig. 6.15 Moderately and severely nuclear heterogeneous mesothelial cells (Ritter-Giemsa stain, 1000×)
The nucleo-cytoplasmic ratio is markedly enlarged and nucleolus is visible; the cells are markedly darkly stained and morphologically intersect with mesothelioma cells. Some of the severely heterogeneous mesothelial cells may be tumor cells, but immunohistochemistry is needed to confirm this, and the morphology is not clear. A large number of fungi (red arrows) can be seen, which, in combination with the patient's history of severe renal disease, long history of abdominal dialysis, and decreased immunity, suggests a fungal infection of the abdominal cavity

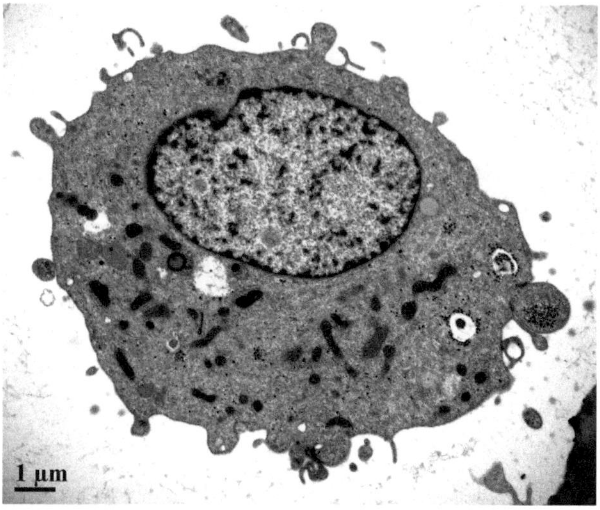

Fig. 6.16 Malignant mesothelioma cells (TM, 12,000×)
The nucleus is predominantly euchromatin; the cell body is mildly enlarged; the cytoplasm has a marked increase in organelles, more mitochondria, dark staining, and pronounced cristae; the increase in free ribosomes and the appearance of glycogen in cluster suggest that the cell synthesis and metabolism are vigorous, making the cytoplasm very darkly stained

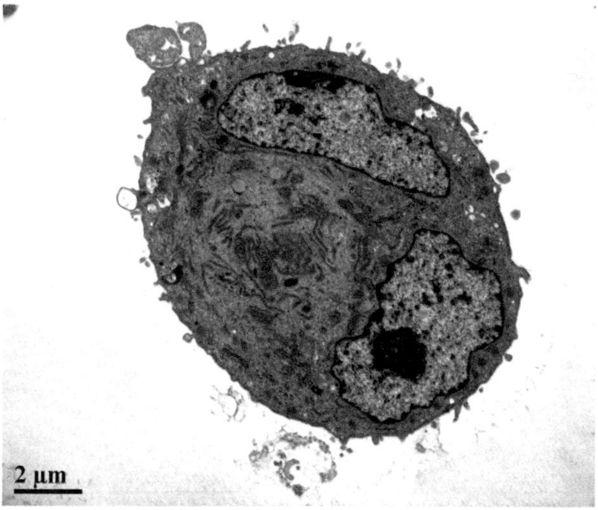

Fig. 6.17 Dinucleated malignant mesothelioma cells (TM, 10 000×)
The nucleus is malformed, euchromatin is dominant, and nucleolus is obvious; the endoplasmic reticulum in the cytoplasm is obviously increased, and the arrangement is disordered, suggesting that the cell synthesis is powerful; the cytoplasm has many free ribosomes and glycogen, and the cell as a whole is grayish-black

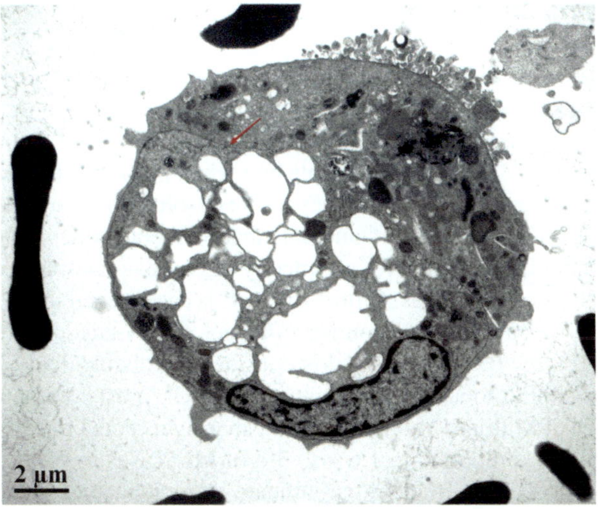

Fig. 6.18 Tumor cell phagosomes (TM, 8000×)
Tumor cells formed phagosomes, and the vacuoles of mesothelial cells were degenerated; microvilli on the outer side of the cell membrane appeared in clusters and densely packed, which were different from the pseudopods and folds of warty protrusions of normal mesothelial cells. A clear plasma membrane boundary (arrow) is seen in this phagosome, suggesting two tumor cells

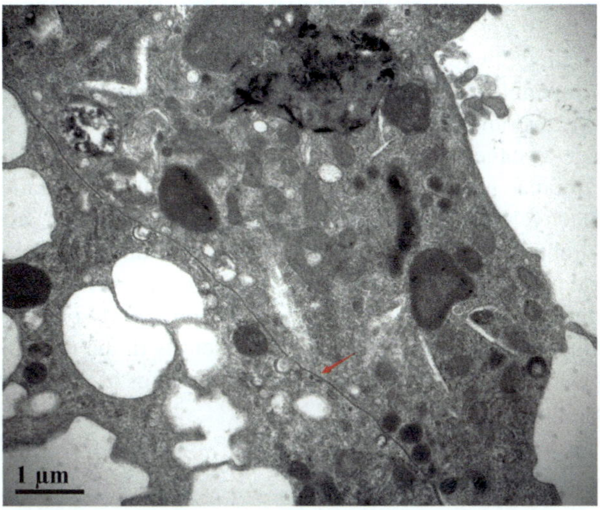

Fig. 6.19 Local magnification of tumor cell phagosomes (TM, 40 000×)

6.4 Plasma Cell Leukemia

Plasma Cell Leukemia (PCL) is a malignant neoplasm of plasma cell origin, characterized by the clonal proliferation of abnormal plasma cells in the bone marrow, infiltration of various organs and tissues, and their substantial presence in the peripheral blood. The diagnosis is established when plasma cells account for more than 20% of peripheral blood leukocytes or the absolute plasma cell count exceeds 2.0 × 10^9/L. Epidemiological data indicate that PCL constitutes approximately 1% to 2% of all acute leukemias, with a typically aggressive and rapid clinical course similar to other acute leukemias. Clinically, PCL is categorized into Primary Plasma Cell Leukemia (PPCL) and Secondary Plasma Cell Leukemia (SPCL). PPCL accounts for 60% to 70% of cases and is considered a distinct leukemia entity with manifestations similar to acute leukemia, while SPCL most commonly develops as a terminal stage of Multiple Myeloma (MM), sharing its core clinical and pathological features. SPCL is reported in 1.6% to 2% of MM cases in Western populations, whereas domestic reports from China indicate an incidence of approximately 8%. Rarely, SPCL may arise secondary to Waldenström's macroglobulinemia, lymphoma, chronic leukemia, or amyloidosis. In the context of diagnostic cytomorphology of body fluid specimens, PCL is relatively uncommon; however, tumor cells corresponding to the primary malignancy are often observed in the body fluids of most patients with metastatic disease.

6.4.1 Plasma Cell Leukemia Pleural Effusion Metastasis

Case: Elderly patient, male, 76 years old, with chronic disease. He was admitted to the hospital with "multiple myeloma diagnosed more than 1 year ago, with fatigue, poor appetite and abdominal pain for 2 weeks." Current diagnosis: multiple myeloma IgA-λ ISS stage I, stage A; secondary plasma cell leukemia; sepsis; community-acquired pneumonia, non-severe; uncompensated acute heart failure; atrial presystole (atrial tachycardia); type 2 diabetes mellitus; urinary tract infection; chronic pancreatitis; lumbar degeneration of the intervertebral discs; pathologic fracture of the T6 vertebral body, which could not be elsewhere categorized; fatty liver; thrombosis of intermuscular veins of the left lower extremity; post-splenectomy; hypernatremia; hypokalemia; hypoproteinemia. The patient is in the terminal phase of the disease and is critically ill. The cytomorphologic findings of the pleural effusion are shown in Figs. 6.20, 6.21, 6.22, and 6.23.

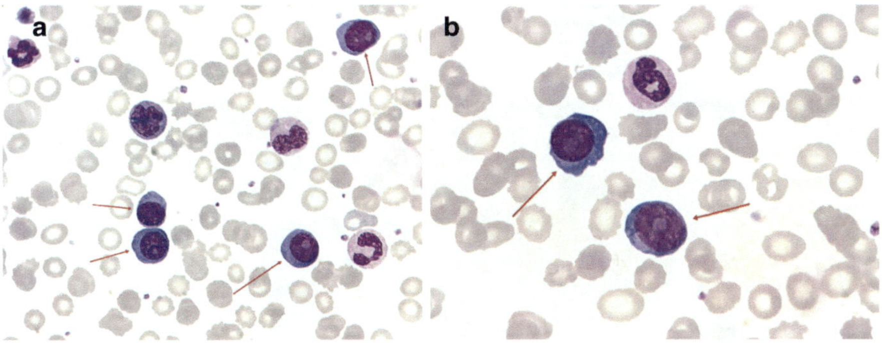

Fig. 6.20 Plasma cell leukemia cells (Ritter-Giemsa stain, 1000×)
Hemorrhagic pleural effusion shows a large number of primitive plasma cells with loose chromatin and one large, conspicuous nucleolus; large nucleo-cytoplasmic ratio, little cytoplasm, perinuclear areas of light staining, cytoplasm blue, few or no granules

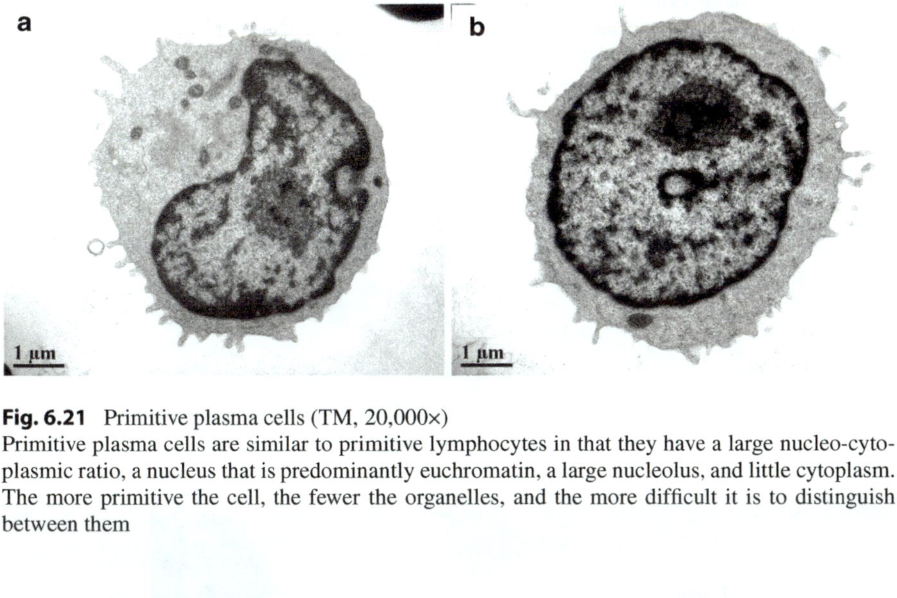

Fig. 6.21 Primitive plasma cells (TM, 20,000×)
Primitive plasma cells are similar to primitive lymphocytes in that they have a large nucleo-cytoplasmic ratio, a nucleus that is predominantly euchromatin, a large nucleolus, and little cytoplasm. The more primitive the cell, the fewer the organelles, and the more difficult it is to distinguish between them

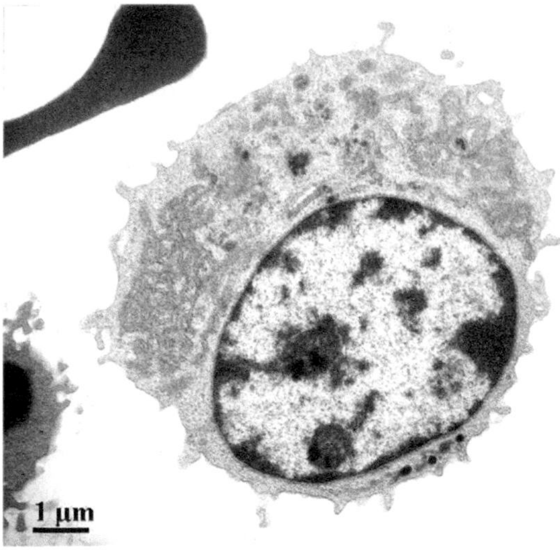

Fig. 6.22 Naïve plasma cells (TM, 15,000×)
The nucleus is predominantly euchromatin, with a large nucleolus; the endoplasmic reticulum is well developed in the cytoplasm, but not like mature plasma cells in a circular, lamellar distribution; Golgi bodies and mitochondria can be seen in the perinuclear region; there is a marked increase in organelles compared with that of primitive plasma cells

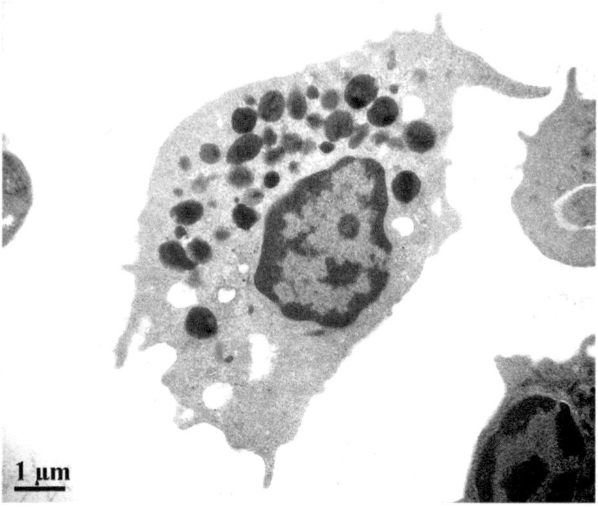

Fig. 6.23 Mature basophils (TM, 15,000×)
Eosinophils and basophils are common in hemorrhagic pleural effusions, mainly due to blood or airborne irritation, which causes blood eosinophils and basophils to free into the tissue fluid and participate in cellular regulation. Mature basophils and eosinophils both have black granules on electron microscopy, but they are slightly different: basophils form white vacuoles because some of the granules are dissolved during electron microscopic staining, while eosinophil granules contain prismatic crystals

6.5 Reactive Plasma Cell

Case: Patient, male, 77 years old, acute course. Two weeks ago, he developed abdominal pain and distension with nausea and vomiting without any obvious cause. Upon admission, he was found to have full abdominal muscle tension, accompanied by tenderness and rebound pain, suspicious mobile turbidities, and weak bowel sounds. CT scan showed a dense shadow of free gas and fluid in the subperitoneum and abdominal cavity. Cavity viscera perforation was considered, so exploratory laparotomy + small bowel perforation repair + appendectomy + abdominal irrigation and drainage were performed under general anesthesia in emergency. After the operation, 1 g of cephalosporin was given twice a day as anti-infection treatment. Six days ago, the patient developed high fever and abdominal distension, and the general bacterial culture result (pus) showed *Escherichia coli*, so the antibiotic was changed to TIENAM 0.5 g once every 8 hours as anti-infection treatment. Five days ago, fecal matter was induced in the patient's drainage tube, and a repeat abdominal CT examination revealed perforation of the digestive tract, blood accumulation in the right middle and lower abdomen, thickening of the intestinal wall and mesentery, and accumulation of fluid in the abdominal and pelvic cavities. Under general anesthesia, exploratory laparotomy + partial sigmoid colon resection + proximal fistula was performed. During the operation, a large amount of purulent fluid was seen in the abdominal cavity, a large number of dried fecal masses were seen in the lower abdomen, and a 4 cm fistula was seen in the distal part of the sigmoid colon at a distance of about 3 cm near the line of peritoneal reflexes, with a rupture of more than 1/2 of the perimeter of the bowel. After the operation, the patient was transferred to the ICU for further treatment. The results of cytomorphologic examination of bilateral pleural fluid are shown in Figs. 6.24, 6.25, and 6.26.

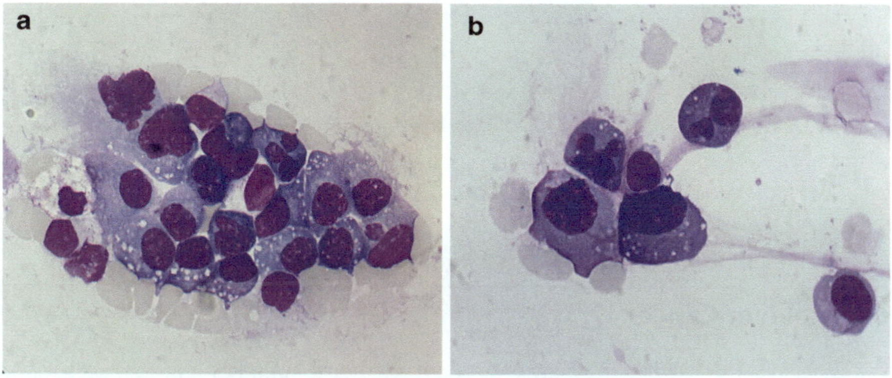

Fig. 6.24 Reactive plasma cells (Ritter-Giemsa Stain, 1000×)
A large number of reactive plasma cells are seen. The cell body is not large, the nucleus is to one side, the chromatin is lumpy, the perinuclear area of pale color is obvious, the cytoplasm is edged with pinkish-purple immunoglobulin, and small vacuoles can be seen in the cytoplasm; a small number of cells are lobulated after nuclear stimulation, and their nuclear morphology changes are similar to that of reactive lymphocytes

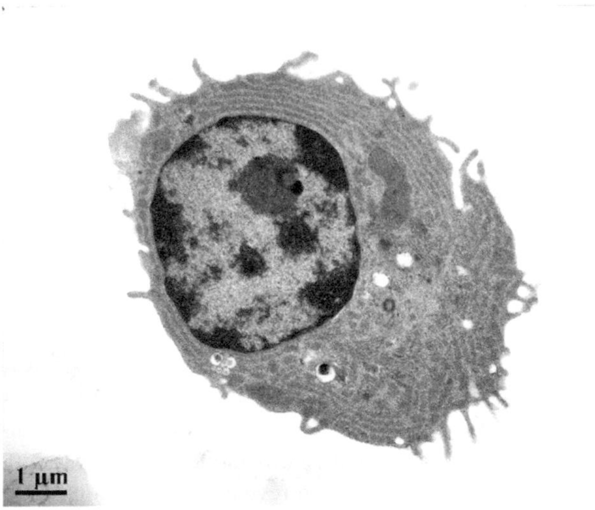

Fig. 6.25 Reactive plasma cells (TM, 15,000×)
Reactive plasma cells are activated mature plasma cells with more heterochromatin in the nucleus and visible nucleolus; cytoplasmic endoplasmic reticulum is well-developed and synthesized; Golgi apparatus and mitochondria are on one side next to the nucleus

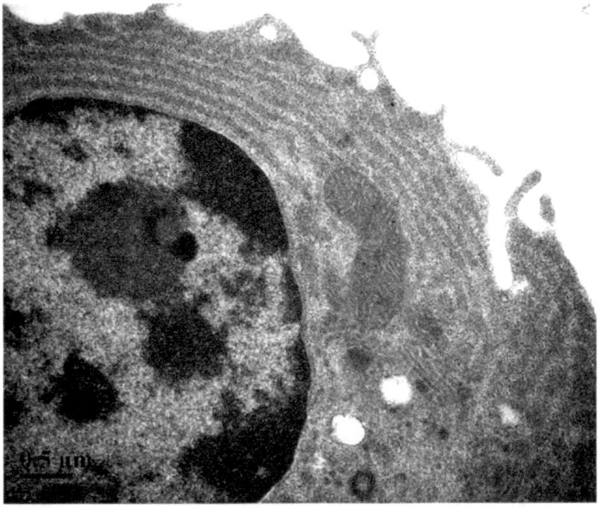

Fig. 6.26 Localized magnified electron micrograph of reactive plasma cells (TM, 30,000×)
The endoplasmic reticulum is seen in the figure arranged in concentric circles in the cytoplasm

6.6 Liver Cancer

Liver cancer, i.e., malignant tumors of the liver, can be divided into two main categories: primary and secondary. Primary liver malignant tumors originate from epithelial or mesenchymal tissues of the liver, and the former is called primary liver cancer, which is a highly prevalent and extremely harmful malignant tumor in China; the latter is called sarcoma, which is less common than primary liver cancer. Secondary or metastatic liver cancer refers to the invasion of malignant tumors originated from several organs of the body to the liver. Generally, liver metastases of malignant tumors of stomach, biliary tract, pancreas, colorectum, ovary, uterus, lung, breast, and other organs are most common. Due to the rich blood vessels in liver, primary liver cancer is very easy to occur systemic blood metastasis, and the main sites of metastasis are lung, brain, bone, and liver cancer cells can also be seen in body fluids. Compared with other tumor cells, tumors originated from liver and other substantial organs have their obvious characteristics in cell morphology.

6.6.1 Primary Liver Cancer Lung Metastasis

Case: Patient, male, 44 years old, acute exacerbation of chronic disease. He was admitted to the hospital with a dry cough for more than 2 months, accompanied by hemoptysis for 1 week. Physical examination: mental clarity, no yellowing of the skin and sclera, clear respiratory sounds in both lungs, and no obvious dry or wet rales were detected. The heart rhythm was unchanged, and no obvious pathologic murmur was detected. The abdomen was soft, with no pressure pain and no rebound pain. Murphy's sign was negative (−), liver and spleen were not touched under the ribs, no obvious masses were felt, shifting dullness was negative (−), and there was no edema of the lower limbs. Diagnoses on admission: hemoptysis; liver transplant status; hepatic malignancy; secondary malignancy of the lungs. The cytomorphologic findings of the pleural effusion are shown in Figs. 6.27, 6.28, 6.29, 6.30, 6.31, 6.32, 6.33, and 6.34.

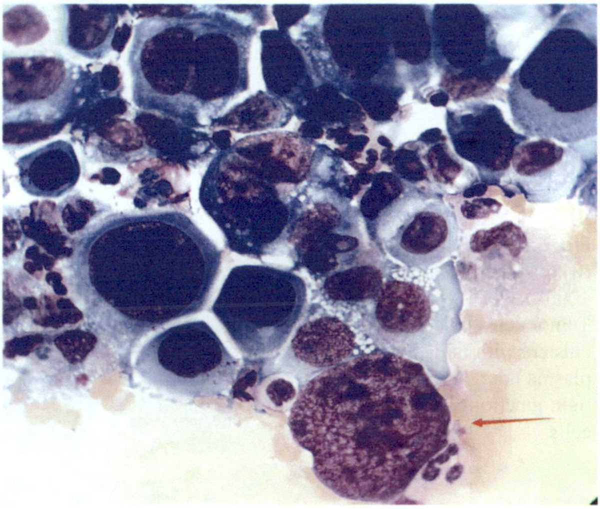

Fig. 6.27 Nests of hepatocellular carcinoma cells (Ritter-Giemsa stain, 1000×)
Liver cancer cells are large in size, with deeply stained nuclei, obvious clefts, and multiple nucleolus; the cytoplasm is blue, and there are few or no acinus. The arrows point to tumor cells with apoptotic naked nuclei, large, misshapen nuclei, multiple nucleolus, and obvious morphological characteristics of tumor cells

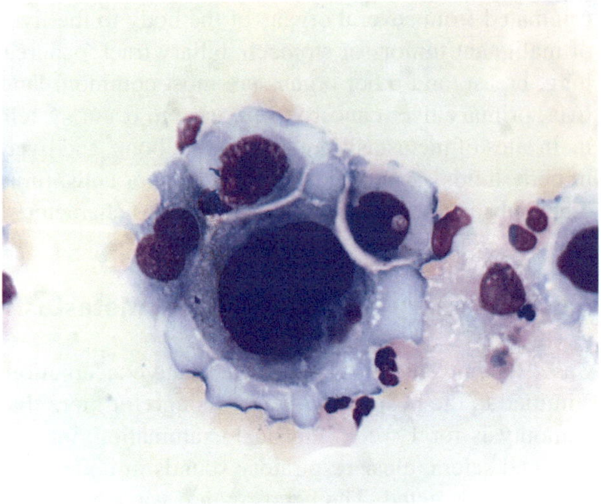

Fig. 6.28 Liver cancer cells (Ritter-Giemsa Stain, 1000×)
The cells were relatively scattered, and the intact liver cancer cells were clearly visible, with large nuclei, obvious clefts, and abundant outer plasma

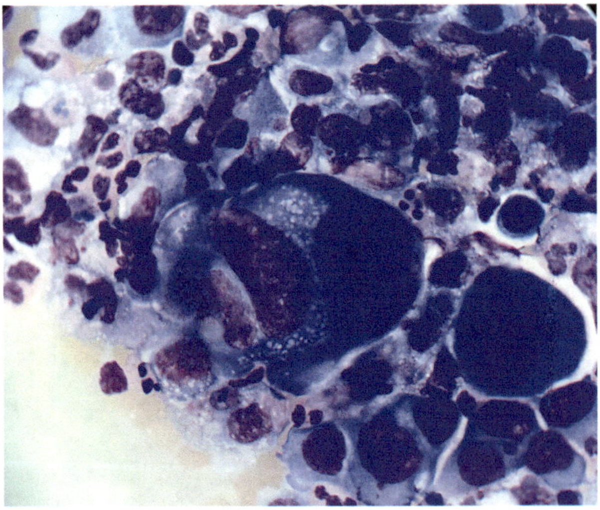

Fig. 6.29 Tumor cell phagosomes (Ritter-Giemsa Stain, 1000×)
Commonly disturbed arrangement of tumor cells, phagocytosis easily seen

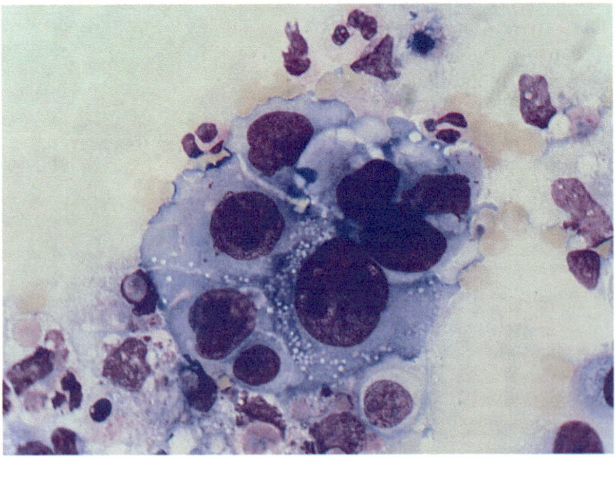

Fig. 6.30 Carcinoma nest (Ritter-Giemsa stain, 1000×)
Tumor cells appear in clusters with indistinct plasma boundaries and tight junctions between cells

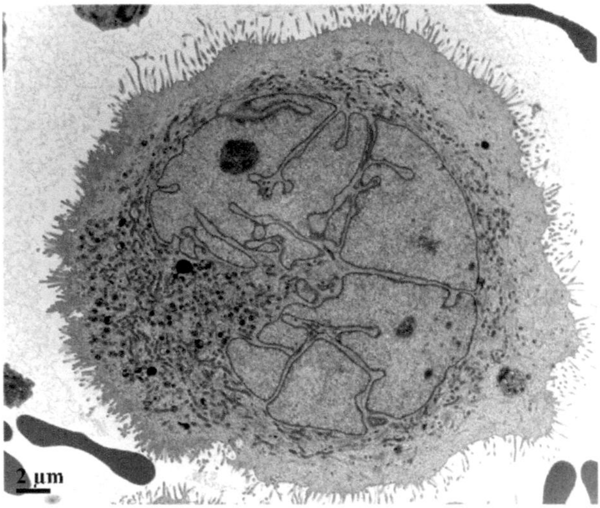

Fig. 6.31 Liver cancer cells (TM, 5000×)
Under the electron microscope, the nuclei of liver cancer cells were obviously fragmented, euchromatin was dominant, and the nucleolus was large; the cytoplasm was rich in organelles, mainly mitochondria; the cellular microvilli were dense and abundant, and under the light microscope, these microvilli behaved as the outer plasma

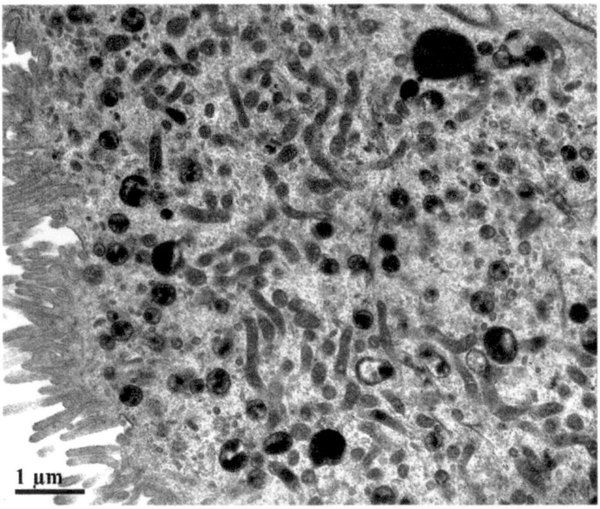

Fig. 6.32 Localized magnified electron microscope image (TM, 40,000×). A large number of mitochondria, more phagocytic vesicles, abundant free ribosomes, and dense microvilli are visible

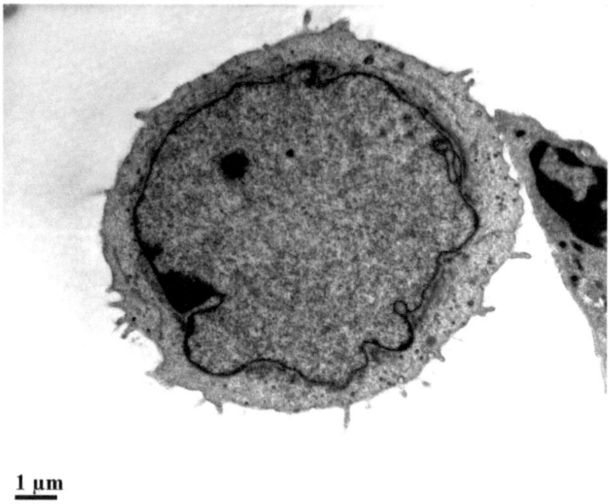

Fig. 6.33 Naïve liver cancer cells (TM, 10,000×)
The nucleo-cytoplasmic ratio is large, and the nucleus is predominantly euchromatin with nucleolus; the amount of cytoplasm is small, and there are few organelles; there are few microvilli. The high number of such cells in the tumor suggests vigorous proliferation

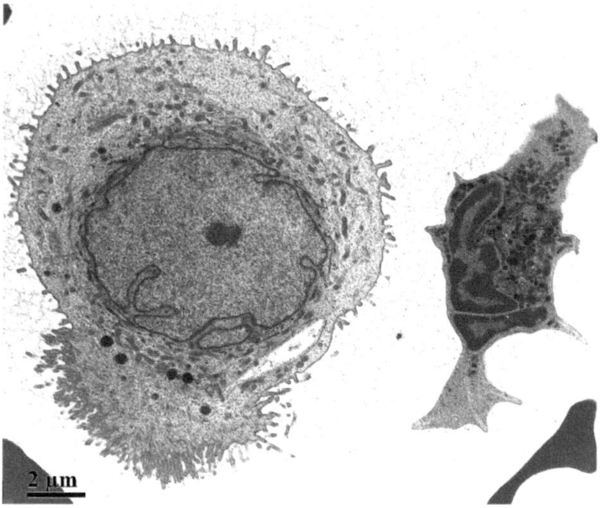

Fig. 6.34 Partially developed microvilli (TM, 8000×)
Some of the microvilli are well developed and appear in clusters, which correlates with their strong transfer and colonization ability

6.7 Dust Cells Increase

6.7.1 Increased Dust Cells in bronchoalveolar Lavage Fluid

Case: Patient, male, 46 years old, with chronic disease. More than ten days ago, a chest CT scan obtained during a health examination reported the following findings: a solid nodule measuring 16 mm × 12 mm, exhibiting well-defined borders, spiculation, and lobulation, was identified in the anteromedial basal segment of the left lower lobe. Multiple small, ill-defined speckled and patchy opacities were present in the left upper lobe. Several small calcifications were observed in both lungs. The lung markings were clear, the thorax was symmetrical, and no other significant abnormal densities were noted. A subsequent contrast-enhanced chest CT scan suggested the following: the solid nodule in the left lower lobe is likely benign, with tuberculosis being the primary consideration; clinical correlation is advised. The opacities in the left upper lobe are indicative of an infectious process, also suspicious for tuberculosis, and further investigation is recommended. The patient was admitted to the hospital with a presumptive outpatient diagnosis of pulmonary nodule. He has a 30-year history of smoking. The cytomorphological findings of the bronchoalveolar lavage fluid are illustrated in Figs. 6.35, 6.36, and 6.37.

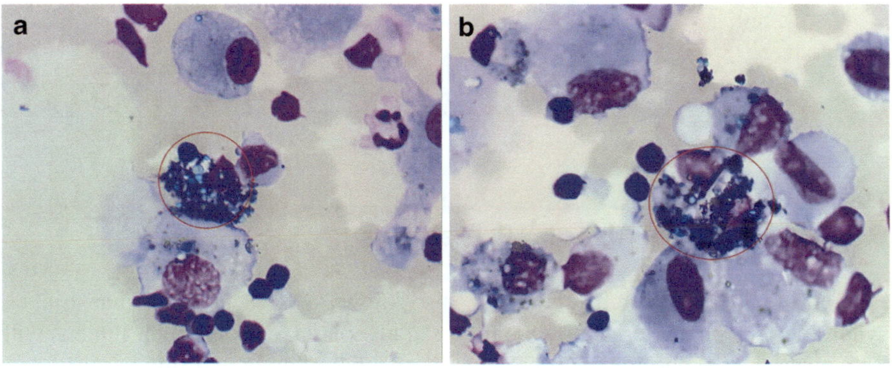

Fig. 6.35 Dust cells (Ritter-Giemsa stain, 1000×)
Within the macrophage, phagocytosed masses of dust form black particles and phagocytosed tar forms pale blue lipid droplets

Fig. 6.36 Dust cells (TM, 10,000×)
Under the electron microscope, the black particles phagocytosed by dust cells are dense, and the lipid droplets are readily solubilized into vacuoles

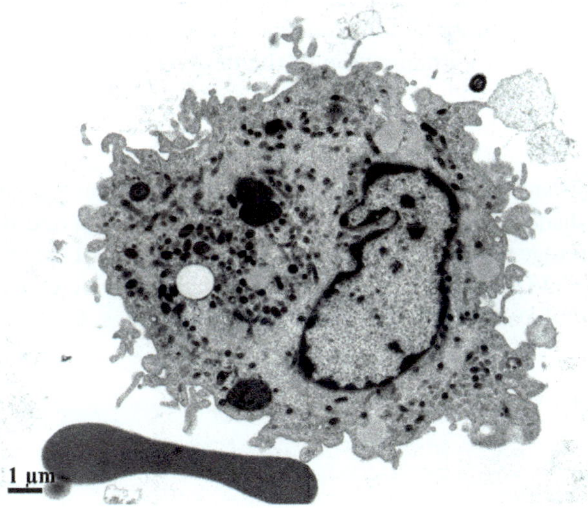

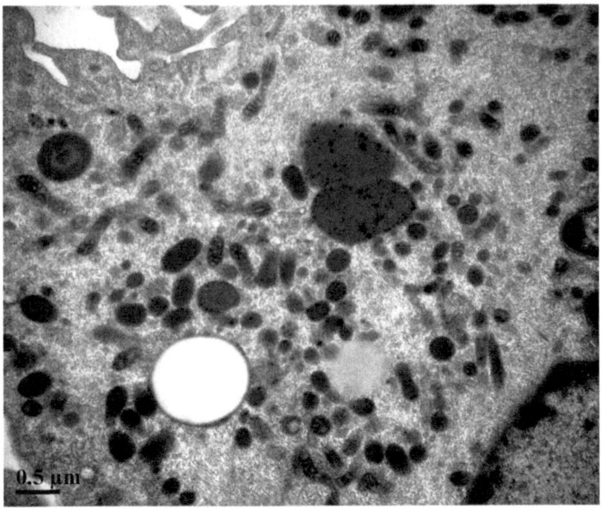

Fig. 6.37 Dust cells (TM, 25,000×)
Macrophage cytoplasm full of dust particles, dense, unlike other inclusion bodies, dust particles are not easily solubilized, irregular morphology

6.8 Pulmonary Alveolar Proteinosis

Pulmonary alveolar proteinosis (PAP), also known as Rosen-Castle-man-Liebow syndrome. It is characterized by the filling of alveolar and bronchiolar lumens with PAS-positive phospholipid-rich proteins. It occurs in young and middle-aged people and is three times more common in men than in women. The cause is unknown and may be related to immune dysfunction. PAP can be caused by dust, especially silica dust, and is thought to be a nonspecific response to certain irritants, leading alveolar macrophages to break down and produce PAS-positive proteins.

Case: Patient, male, 37 years old, with chronic disease. One year ago, he went to the local hospital for examination because of a foreign body sensation in his throat, and the CT examination of the lungs showed scattered flaky ground-glass density shadows with blurred edges in the basal and posterior segments of the lower lobes of both lungs. There was no cough, sputum, dizziness, headache, nausea, vomiting and other discomforts, and no treatment was given. In order to seek further diagnosis and treatment, the patient was admitted to the outpatient clinic as a case of "lung shadow." Routine histopathologic examination and diagnosis (left lingual branch lavage mucosa) showed that "special staining of the left lingual branch lavage fluid" could be consistent with pulmonary alveolar proteinosis. The morphologic findings of the alveolar lavage fluid are shown in Figs. 6.38, 6.39, 6.40, and 6.41.

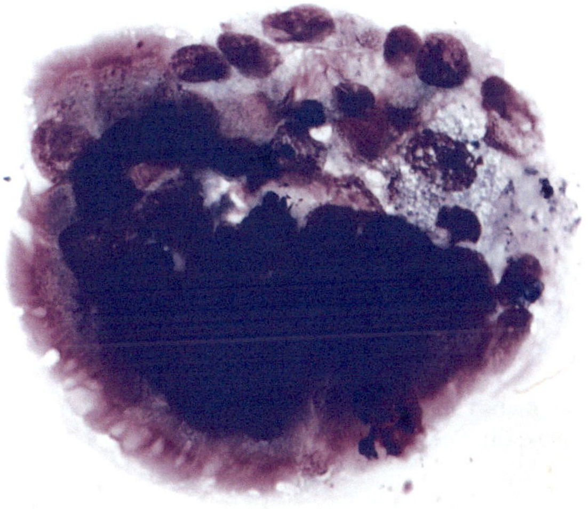

Fig. 6.38 Clusters of ciliated columnar epithelial cells (Ritter-Giemsa Stain, 1000×)

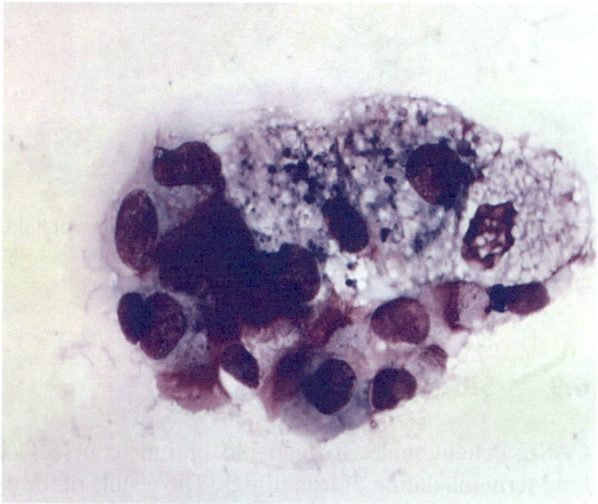

Fig. 6.39 Vacuolar changes in macrophages due to phagocytosis of lipids (Ritter-Giemsa stain, 1000×)

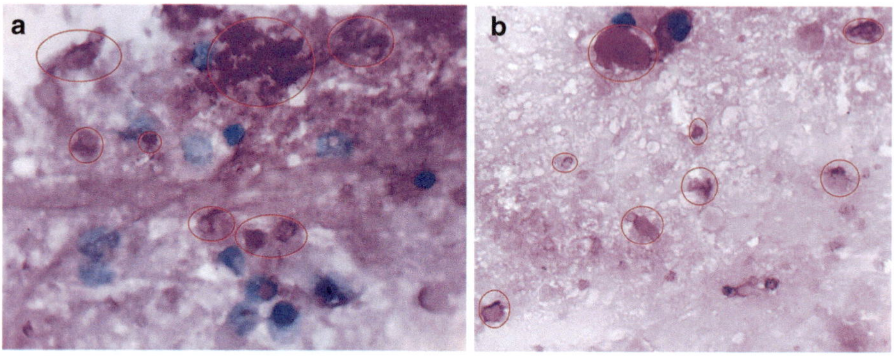

Fig. 6.40 Large number of pinkish-purple-red particles with no fixed shape ((**a**) PAS stain, 1000×; (**b**) PAS stain, 400×)

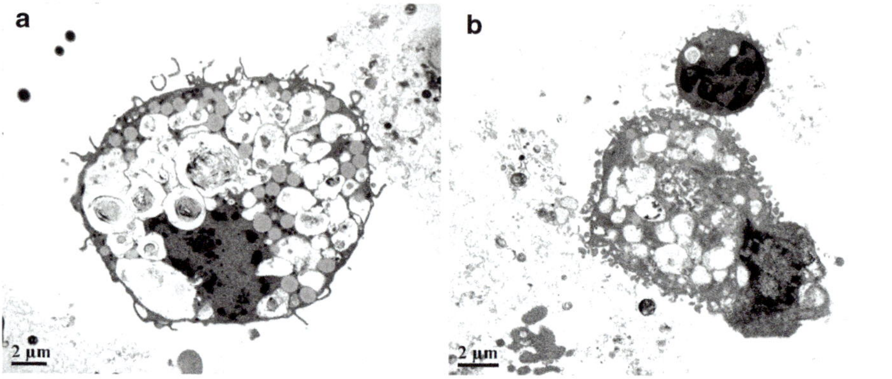

Fig. 6.41 Macrophages that phagocytose large amounts of lipids ((**a**) TM, 7000×; (**b**) TM, 8000×) Some of these lipids are digested, forming residues, so that such macrophage foam-like changes are evident

6.9 Siderosis

CASE: Patient, male, 46 years old, chronic course. Occupational disease caused by long-term inhalation of iron filings. The results of the cytomorphologic examination of body fluids are shown in Figs. 6.42, 6.43, 6.44, and 6.45.

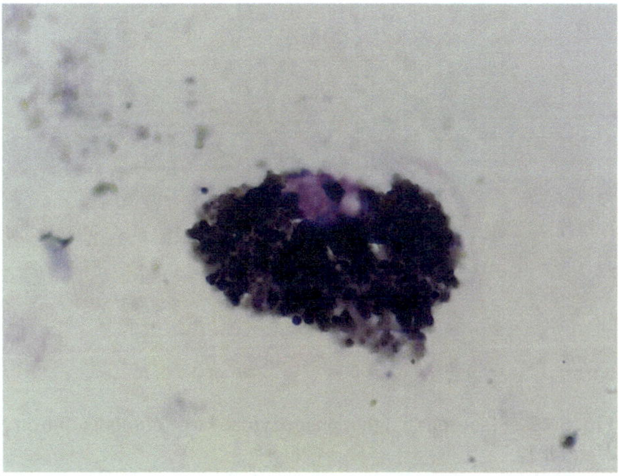

Fig. 6.42 Iron particles in macrophages, black in color, slightly darker than dust cells, not clearly distinguishable by Ritter staining, requires iron staining for confirmation (Ritter-Giemsa stain, 1000×)

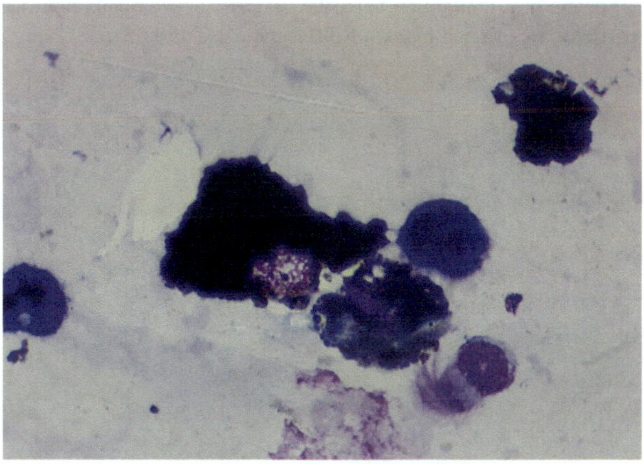

Fig. 6.43 Macrophage cells (Ritter-Giemsa stain, 1000×)
Phagocytosis of a large number of irregularly shaped and sized iron particles that stain positively for iron and are blue in color

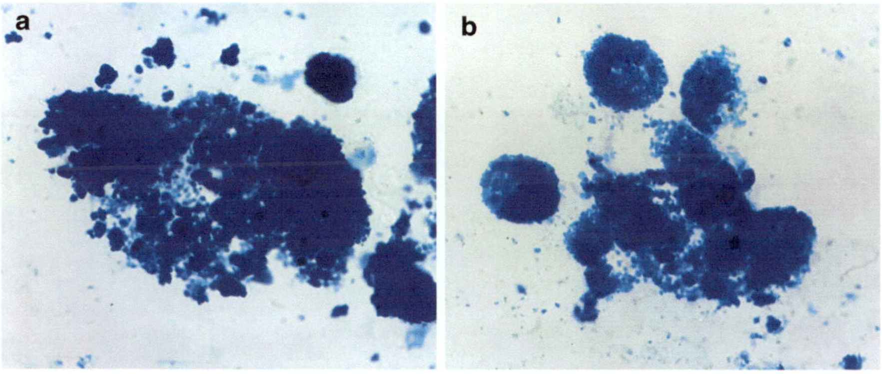

Fig. 6.44 Iron particles engulfed in macrophages, black in color, not clearly distinguishable by Ritter staining, iron staining required for confirmation (iron stain, 1000×)

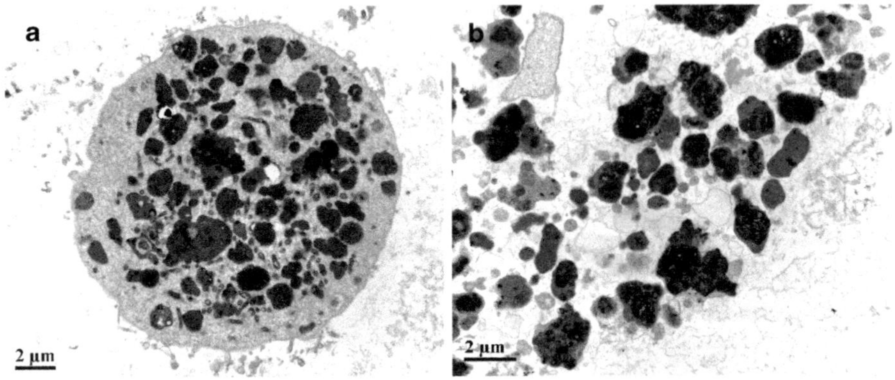

Fig. 6.45 Macrophages that phagocytose large numbers of iron particles ((**a**) TM, 8000×; (**b**) TM, 10 000×)
Figure (**b**) shows a partial enlargement of Figure (**a**). Compared with the iron-containing hemoflavin cells, the macrophages that phagocytose iron particles have black, more irregular particles, while the iron-containing hemoflavin particles are black, homogeneous, and rounder; the dust cell particles are grayish-black, of different sizes, and relatively tiny

Light and Electron Microscopy Comparison of Microorganisms and Crystallization

7

Dayong Gu and Ying Wang

7.1 Fungi

Fungal infections are commonly associated with gastrointestinal tract perforation, repeated inappropriate puncture procedures (e.g., peritoneal dialysis, chest drainage, etc.), or poor catheter maintenance and are usually associated with severe infections with other pathogens. Because of the large number of disorganized substances in the fluid, which makes diagnosis difficult, it is important to be able to better identify these pathogens in order to provide direction for clinical diagnosis and treatment. The following is a list of the morphology of some common substances in the effusion in electron microscopy and compared with the light microscope, aiming to provide a new means for you to improve the ability of morphologic analysis. See Figs. 7.1, 7.2, 7.3, and 7.4.

D. Gu (✉) · Y. Wang
Laboratory Department, Shenzhen Second People's Hospital, Shenzhen, Guangdong, China

D. Gu, Y. Wang (eds.), *Morphological Diagnosis of Humoral Cells under Light and Electron Microscopy*, https://doi.org/10.1007/978-981-95-8668-4_7

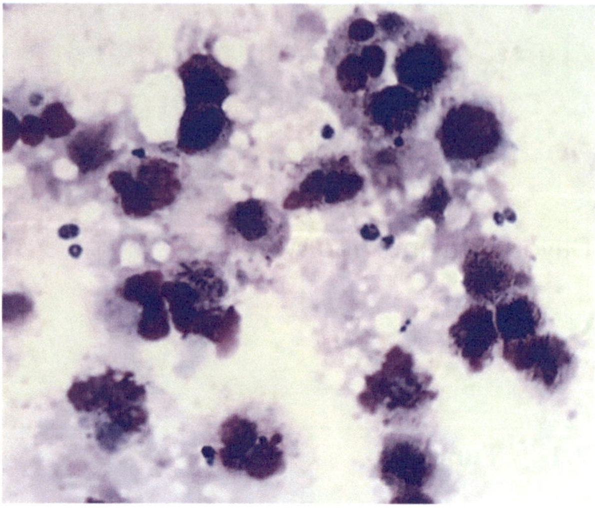

Fig. 7.1 Esophageal pleural fistula. Fungus and apoptotic cells are seen in the purulent effusion (Ritter-Giemsa stain, 1000×)

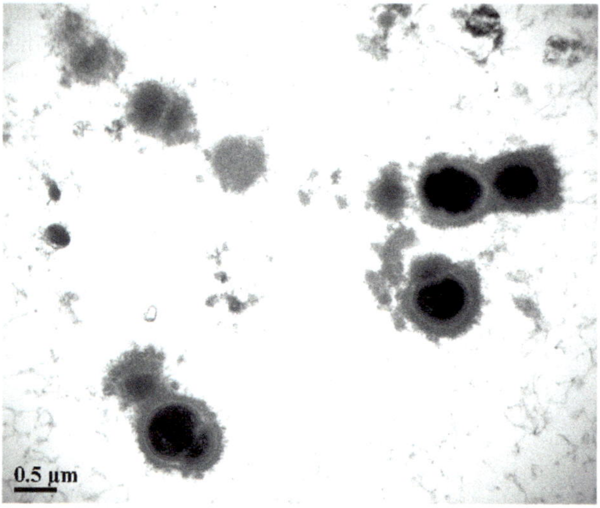

Fig. 7.2 Fungi (TM, 25,000×)

The cell wall of fungi is thicker, concentrating about 30% of the dry matter of the cell, which can maintain the shape of the cell and resist external influences, and is mainly composed of hexose or aminocaprose consisting of polysaccharide chains β1,3-β1,6-glucan. The cell wall of most fungi consists of chitin, which is very tough, unlike the cell wall of plants, which is mostly composed of cellulose, the cell wall of low-grade fungi is also cellulose-based, and the cell wall of yeast is mainly composed of glucan-based, the cell wall of different classes of fungi has different components, and its composition is closely related to the classification of the fungus and its resistance

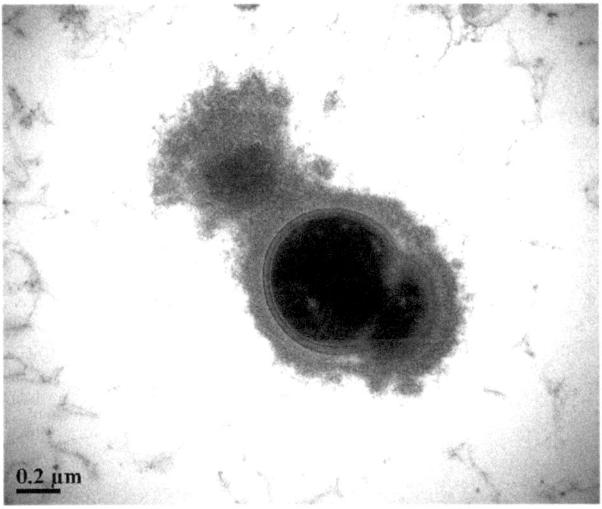

Fig. 7.3 Fungi (TM, 60,000×)

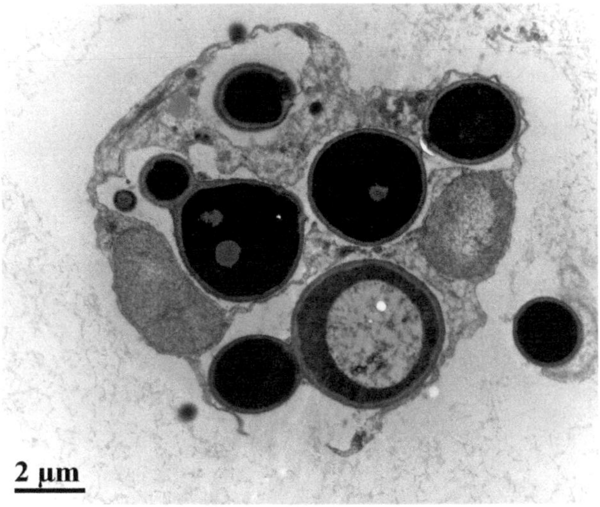

Fig. 7.4 Intracellular Fungi (TM, 10 000×) After the fungus is engulfed by the cell, part of the fungus is digested, and the structure is not clear

7.2 Viruses

A variety of bacteria are commonly found in infected fluid. Since the cytomorphology of body fluids is based on the Ritter-Giemsa stain, which can only roughly differentiate between cocci and bacilli, Gram stain or other special stains are required, as well as the combination of bacterial culture and identification, in order to clarify the type of infection. See Figs. 7.5, 7.6, and 7.7.

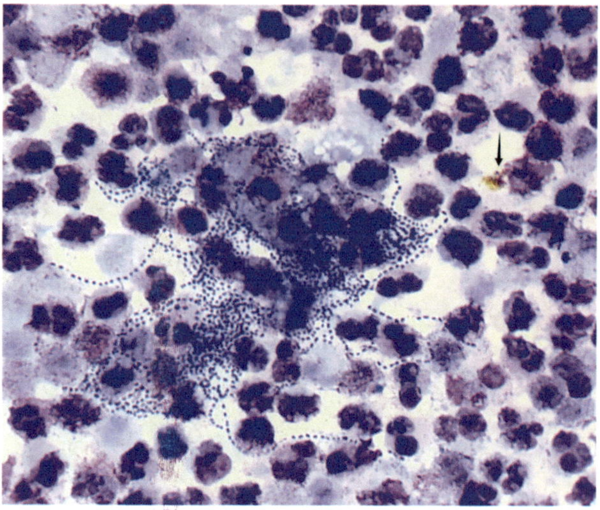

Fig. 7.5 Esophageal pleural fistula, large number of cocci in purulent pleural effusion, bilirubin crystals (arrows), suggesting old hemorrhage (Ritter-Giemsa stain, 1000×)

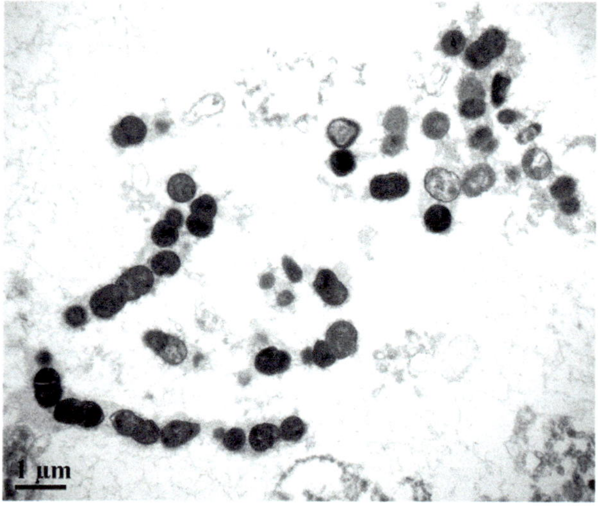

Fig. 7.6 Strings of cocci (TM, 15 000×)

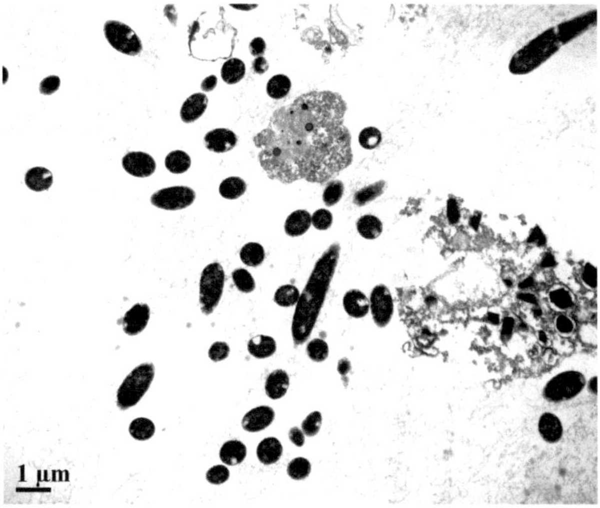

Fig. 7.7 Coccidia and bacilli (TM, 15 000×) with different cut sections

7.3 Charcot-Leyden crystallization

Charcot-Leyden crystals are rhombic, colorless, transparent, and compass-like, with pointed ends, unequal sizes, and strong refractive properties. They are formed when eosinophilic particles fuse with each other after the rupture of eosinophilic granulocytes and belong to the protein crystals, which are most commonly found in parasitic infections such as *Schistosoma pneumoniae*, amoebic dysentery, and hookworm disease, as well as in allergic diseases and hematochezia.

Case: Male, 24 years old, acute course. One month ago, after eating raw shrimp, he developed chest tightness and shortness of breath, unable to lie down, which was obvious after activity and slightly relieved after resting, and was examined in the local hospital in Anhui Province, China, and found the following: Echinococcus, IgG antibody-positive; *Paragonimus westermani*, IgG antibody-positive; *Schistosoma japonicum* Katsurada, IgG antibody-positive; *Fasciola hepatica* Linnaeus westermani, IgG antibody-positive; and sparganum: IgG antibody-positive. Blood sedimentation and CRP are elevated. Chest CT showed infectious lesions in both lungs, pleural effusion on both sides, pericardial effusion, and enlarged mediastinal lymph nodes, and the diagnosis was "schistosomiasis." He was treated with thoracentesis for drainage of fluid and antibiotics (specific drugs and dosage are not known) for anti-infective treatment and was discharged from the hospital. After discharge from the hospital, the patient returned to the outpatient clinic and took oral praziquantel tablets for symptomatic treatment, but he still had chest tightness and discomfort, accompanied by shortness of breath after activity, so he came to our hospital and was admitted to the outpatient clinic as a case of pleural effusion. The cytomorphologic findings in the pericardial effusion are shown in Figs. 7.8 and 7.9.

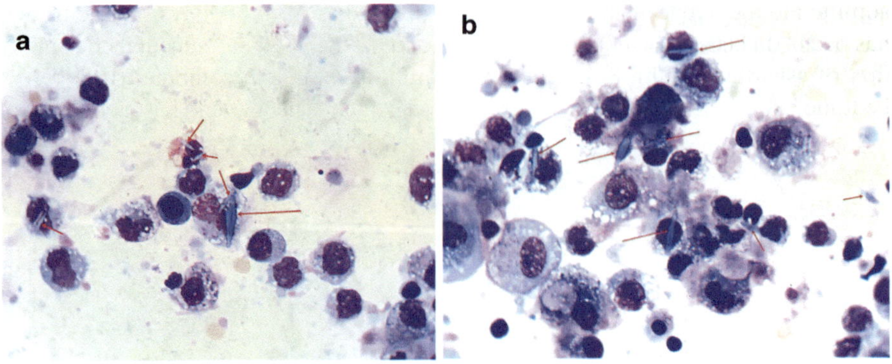

Fig. 7.8 Charcot-Leyden crystals can be seen both intracellularly and extracellularly (Ritter-Giemsa stain, 1000×)

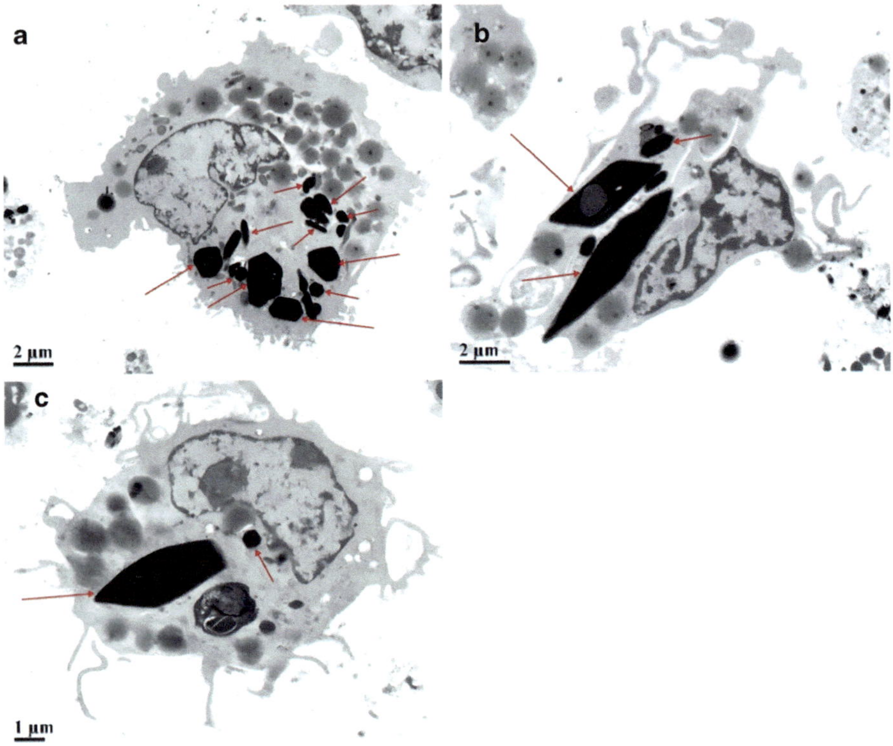

Fig. 7.9 Intracellular Charcot-Leyden Crystals
(**a**) Charcot-Leyden crystal cross-section (TM, 8000 ×), hexagonal or irregular black dense crystals; (**b**) Charcot-Leyden crystal oblique cross-section (TM, 10,000 ×); (**c**) Charcot-Leyden crystals (TM, 12,000 ×), due to electron microscopy specimen preparation need to be embedded, staining process, extracellular crystals are too small, easy to be lost, and only the intracellular crystals are visible. Because the crystals are composed of proteins, they appear densely black in color under the electron microscope, similar to hemoglobin imaging

7.4 Sodium Urate Crystals

Sodium urate crystals are common crystals in joint fluid, causing chronic inflammation and fibrous tissue proliferation to form stones, also known as gout stones. Most of them are scattered colorless needles, or in large quantities, and they can be flocculent, uncolored, and of different sizes and can be phagocytosed by neutrophils and macrophages, as shown in Figs. 7.10, 7.11, and 7.12.

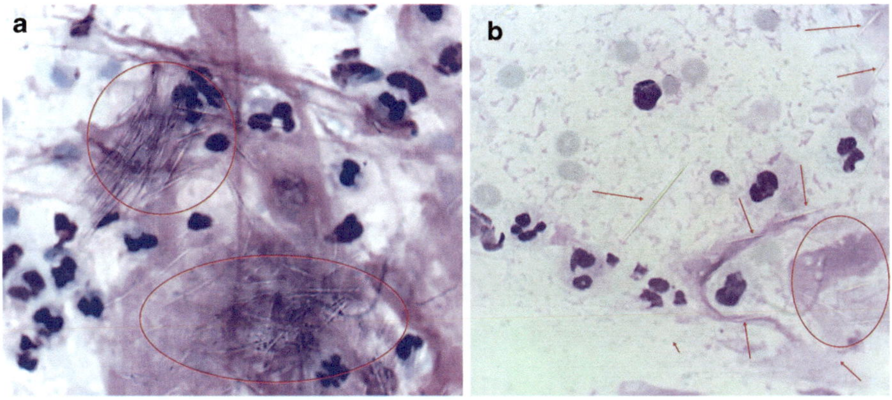

Fig. 7.10 Clusters and individual sodium urate crystals (Ritter-Giemsa stain, 1000×)

Fig. 7.11 Sodium urate crystals phagocytosed by cells (Ritter-Giemsa stain, 1000×)

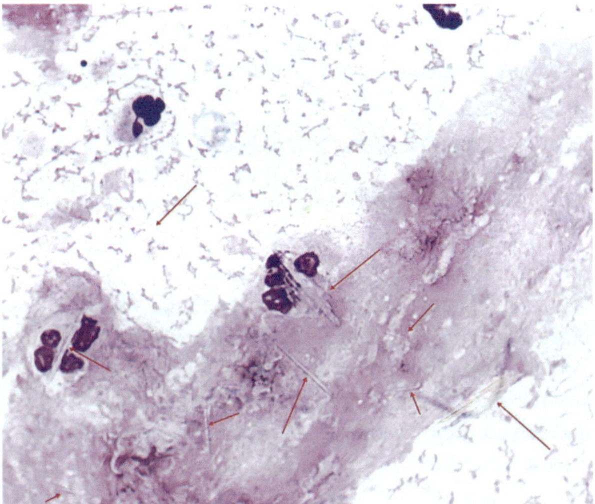

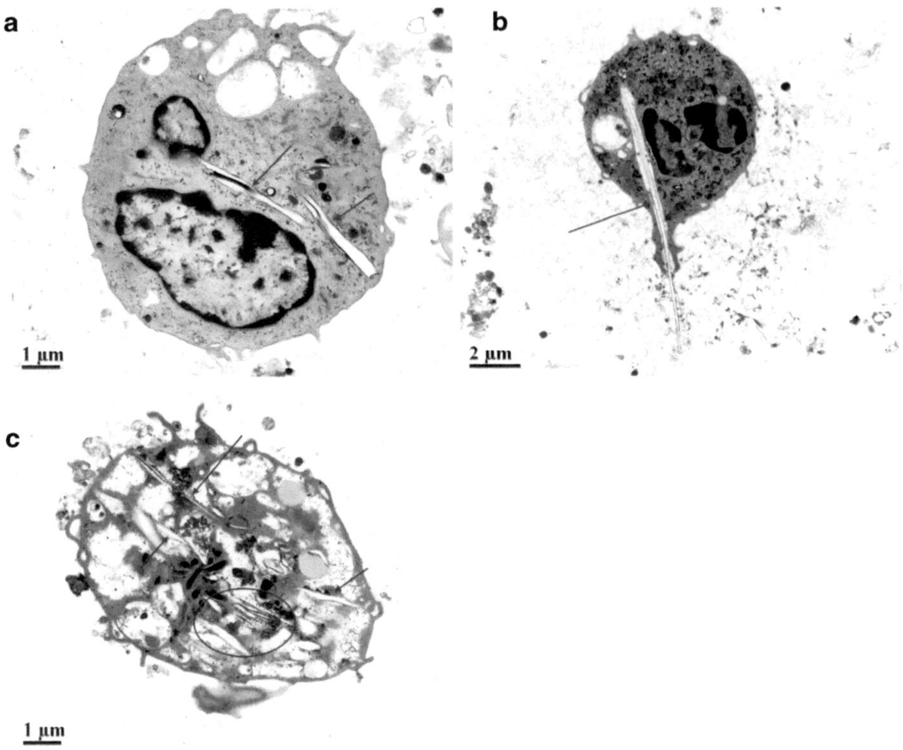

Fig. 7.12 Sodium urate crystals
Under the electron microscope, sodium urate crystals are inorganic crystals, transparent, and needle-like; Charcot-Leyden crystals are protein crystals, black in color, which are completely different (A. TM, 15 000 ×; B. TM, 10 000 ×; C. TM, 15 000 ×)

7.5 *Tropheryma whipplei*

Tropheryma whipplei (TW) is a Gram-positive bacterium, aerobic, 0.25–0.5 μm × 1.5–2.5 μm in size, with a unique three-layer plasma membrane cell wall ultrastructure under electron microscope. The 16S rRNA lineage analysis and 16S–23S spacer sequence revealed that TW belongs to the phylum *Actinobacteria*, class *Actinomycetes*, order *Actinobacteria*, family *Cellulomonadaceae*, and genus *Tropheryma*. Sequence analysis of the whole genome showed that TW has a single small ring chromosome (925938 bp), and sequence analysis showed that it is a defective type of bacteria. It is a symbiotic bacterium. Morphologically, it is a PAS-positive bacillus within the cell bodies of foamy macrophages. TW is a widespread microorganism, but patients with Whipple's disease (WD) are extremely rare. Exposure to TW does not necessarily lead to the development of WD, and certain immunopathologic factors may play an important role in the pathogenesis of WD. The immunodeficiency of WD patients may be an important factor in their susceptibility to TW, as shown in Figs. 7.13, 7.14, 7.15, and 7.16.

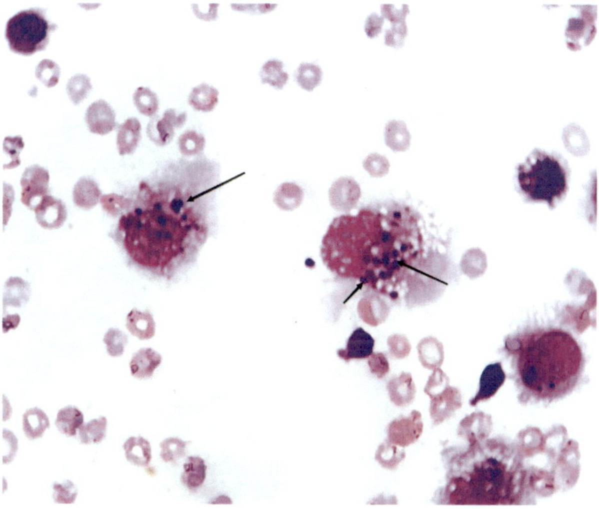

Fig. 7.13 *Tropheryma whipplei* in bronchoalveolar lavage fluid (black arrow, Gram stain, 1000×). In the cytoplasm of foamy macrophages, Gram-positive short bacilli

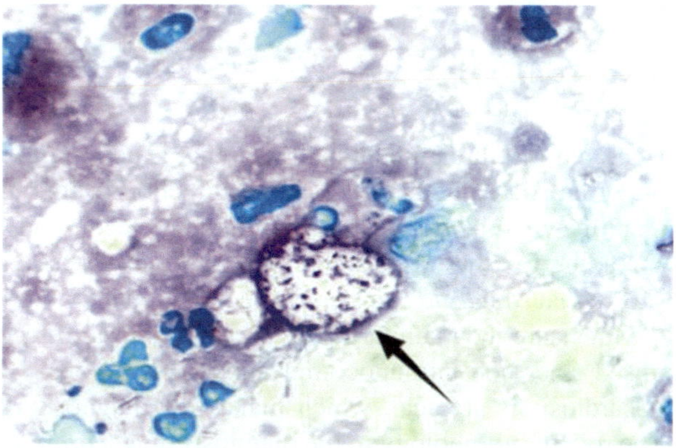

Fig. 7.14 The purplish-red bacteria in foamy macrophages are *Tropheryma whipplei* (black arrow, PAS stain, 1000×)

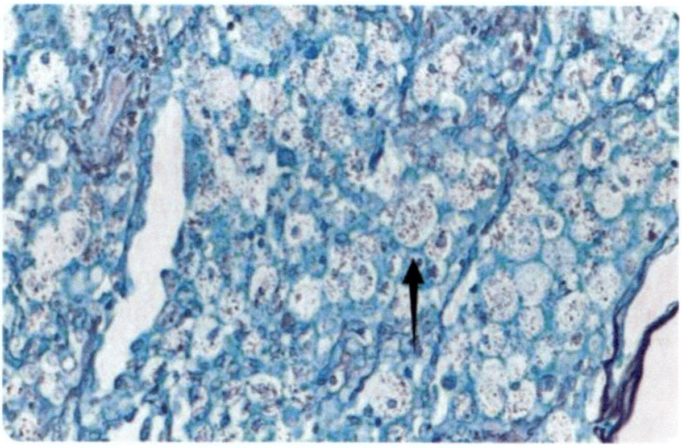

Fig. 7.15 Lung tissue section showing foamy macrophages of varying sizes, intracellular and extracellular positive bacilli (silver hexamine stain, 400×)

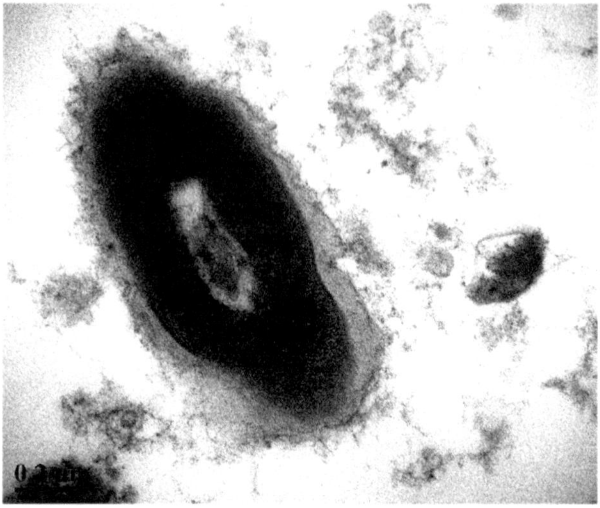

Fig. 7.16 The *Tropheryma whipplei* shows a typical three-layer plasma membrane structure (TM, 100 000×)

7.6 *Nocardia*

Nocardia is a rare clinical pathogen that is ubiquitous in the environment and is an opportunistic pathogen that causes infection in immunocompromised individuals, resulting in suppurative lung infections, ulcers, and multiple fistulas, as well as spreading to other organs and causing brain abscesses and peritonitis. It has been shown that *Nocardia* can also cause infection in immunocompetent patients and that the lungs are the main target organ for *Nocardia* infection in both immunocompromised and non-immunocompromised conditions. The clinical presentation of nocardiosis is atypical, which makes the diagnosis of nocardiosis more difficult. The role of morphology in the diagnosis of *Nocardia* infection: when sputum smear or lavage fluid examination, the specimen is too thick, *Nocardia* are mostly distributed in the pus, it is not easy to color, or cause the staining of the negative and positive uncertainty. *Nocardia* present pairs and clusters of filamentous mycobacteria, mostly interspersed between neutrophils, right-angle branching, and other morphological characteristics. *Nocardia* grows slowly, and the culture is often masked by the dominant bacteria, resulting in *Nocardia* not being easily detected or even missed. Conventional morphological examination is simple and fast and does not require special instruments, and the report will be issued in a short period of time. *Nocardia* has its unique morphological characteristics, and with strong morphological diagnostic ability, it will be able to give hints as early as possible. See Figs. 7.17, 7.18, 7.19, 7.20, and 7.21.

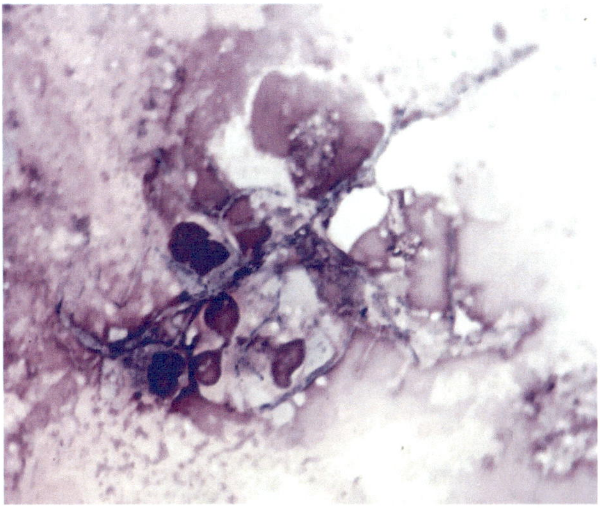

Fig. 7.17 *Nocardia* in bronchoalveolar lavage fluid, not easily stained (Ritter-Giemsa stain, 1000×)

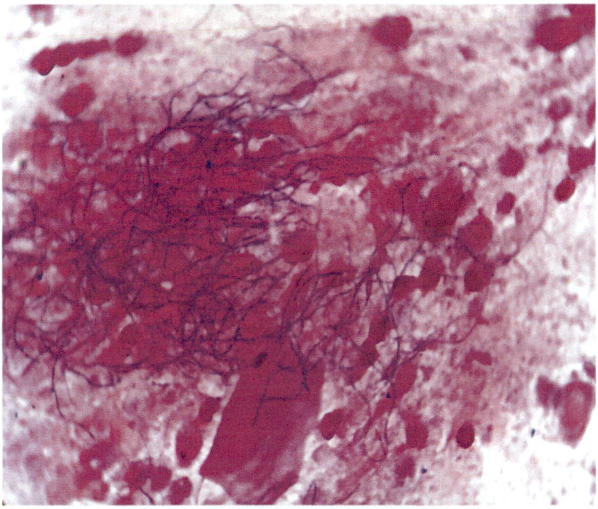

Fig. 7.18 *Nocardia* distributed in pus, disorganized and distributed in clusters (Gram stain positive, 1000×)

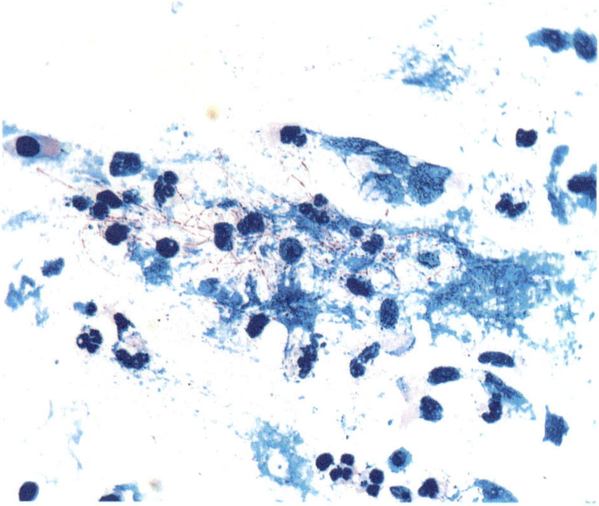

Fig. 7.19 Weakly positive, clustered *Nocardia* (weak antacid stain, 1000×)

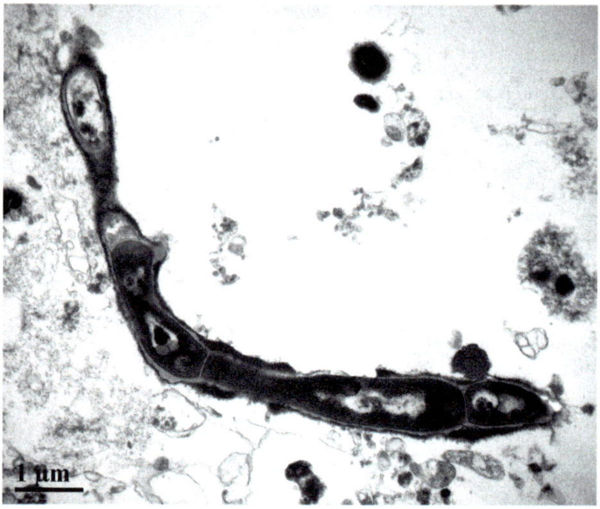

Fig. 7.20 *Nocardia* with distinct cell wall structures and L-shaped branches (TM, 20 000×)

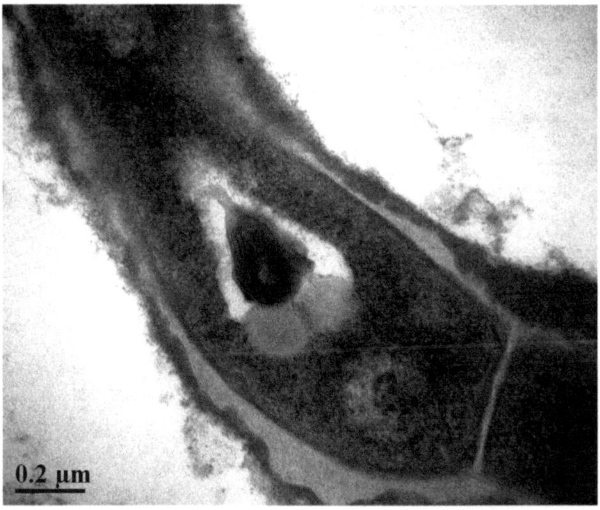

Fig. 7.21 *Nocardia* with visible distinct cell wall structure (TM, 100 000×)

Epilogue

The application of electron microscopy in fluid cell morphology complements the limitations of conventional examinations in recognizing details, playing a crucial role in the diagnosis and differential diagnosis of tumors and complex cases. The publication of this book is just a beginning, and we hope that more morphology experts will join in, making EM morphology richer and more concrete, adding a new perspective to morphology.

The days and nights of silent effort, though everyone worked hard, the exploration process was full of joy. We believe that perseverance will pay off, and what is longed for will eventually come to fruition.

We hope that this book can contribute a bit to the development of morphology and serve as a reference for laboratory departments of medical institutions at all levels, independent laboratories, laboratory technicians, instructors, and clinical doctors.

Due to the editors' limited knowledge and experience, there may be errors or inappropriate content in the book. We kindly ask the readers and experts for criticism and correction, so that it can be supplemented and improved in future editions.

To all the dedicated morphology workers who steadfastly uphold their beliefs.

D. Gu, Y. Wang (eds.), *Morphological Diagnosis of Humoral Cells under Light and Electron Microscopy*, https://doi.org/10.1007/978-981-95-8668-4

Group photo of the editor-in-chief of this book in Zhejiang Provincial People's Hospital for further training

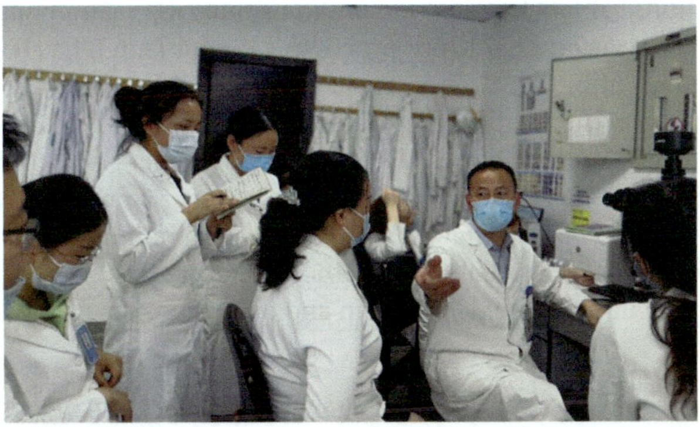

Prof. Wu Mao teaching students

This book was greatly supported during its compilation by the Electron Microscopy Center (Life Sciences Division) at Zhejiang University, for which we are deeply grateful! Over the past 40 years, the Electron Microscopy Center (Life Sciences Division) at Zhejiang University has developed into a first-class biological microstructure analysis platform in China, accumulating rich experience in the ultrastructure of animals, plants, microorganisms, virus morphology, and medical pathology electron microscopy diagnosis. The center has established related techniques such as biological material ultrathin sectioning, cryo-ultrathin sectioning, electron microscopy combined with immunogold labeling, fluorescence localization, electron microscopy cytochemistry, rapid electron microscopy diagnosis, scanning electron microscopy surface structure and composition analysis, and more.

The laboratory has four 120KV transmission electron microscopes, one cold field emission high-resolution scanning electron microscope, three tungsten filament scanning electron microscopes, two scanning electron microscope spectrometers, three laser confocal microscopes, and two fluorescence microscopes and is also equipped with a series of advanced sample pre-processing equipment such as several ambient ultrathin microtomes, frozen ultrathin microtomes, immuno-metallurgical labeling instrument, ion sputtering instrument, critical point dryer, and other advanced sample pre-processing equipment. It has been rated as the demonstration unit of bio-electron microscope in the "Instrumentation and Quality Resource Sharing System of Higher Education Institutions" of the Ministry of Education of China, and has been listed as the "Zhejiang Electron Microscope Center (Life Science Division)" by the Department of Science and Technology of Zhejiang Province. As an important instrument sharing unit in Zhejiang Province, it provides professional, high-quality and efficient services to universities, research institutes, medical units, and enterprises in and out of the province with the concept of open sharing.

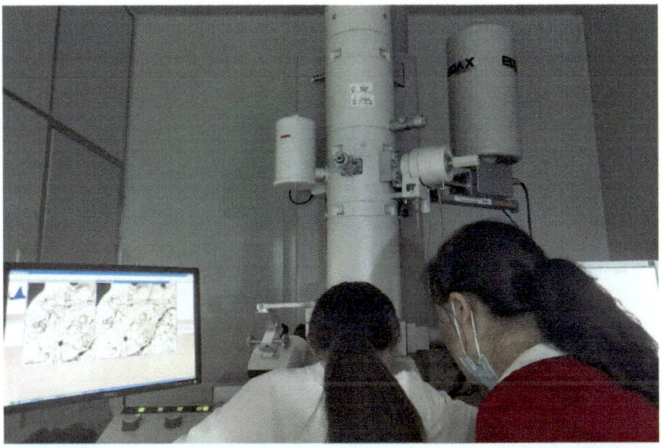

The editor-in-chief of this book is taking electron microscope photos at the Electron Microscope Center (Department of Life Sciences) of Zhejiang University

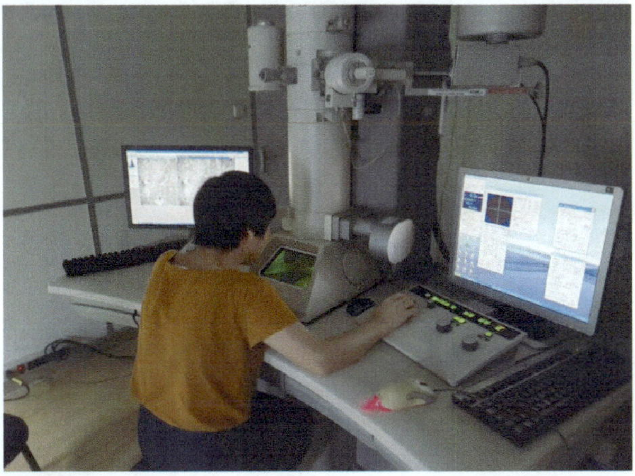

Undergoing ultrastructural observation and analysis

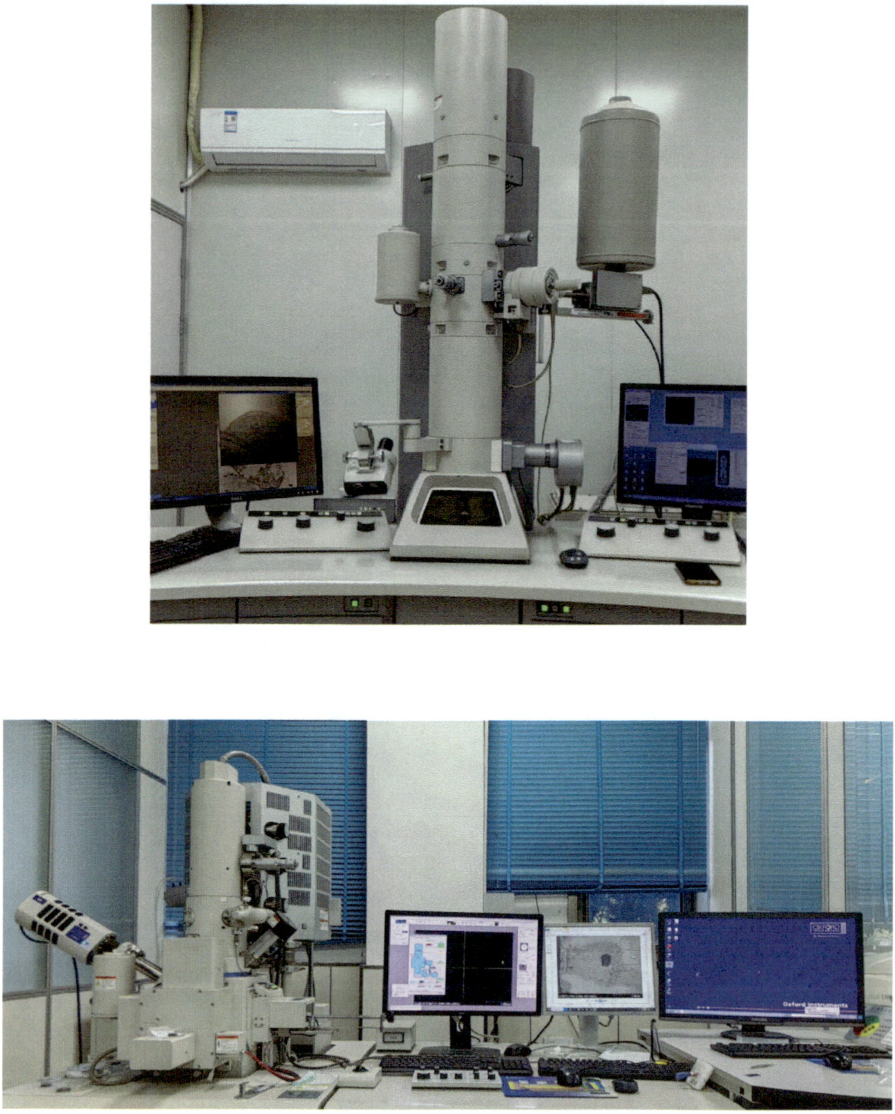

Display of some instruments and equipment in the Microscope Center

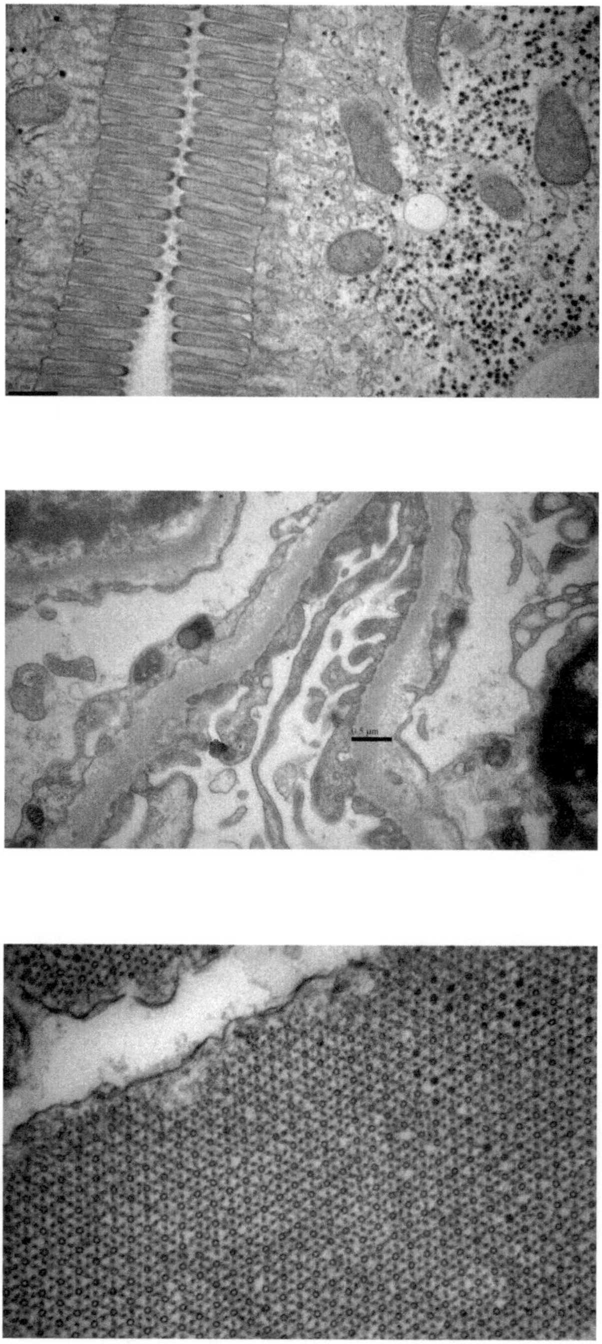

Display of ultrastructure observation and analysis

GPSR Compliance

The European Union's (EU) General Product Safety Regulation (GPSR) is a set of rules that requires consumer products to be safe and our obligations to ensure this.

If you have any concerns about our products, you can contact us on ProductSafety@springernature.com

In case Publisher is established outside the EU, the EU authorized representative is:

Springer Nature Customer Service Center GmbH
Europaplatz 3
69115 Heidelberg, Germany

Batch number: 10732090

Printed by Printforce, the Netherlands